The True Adventures of Jim Stark

Jim Stark

Copyright © 2016 Jim Stark

No part of this book may be reproduced by any means without the written permission of the author.

All rights reserved.

Manufactured in the United States of America

The True Adventures of Jim Stark
Written by
Jim Stark

www.jimstarknavy.com

Published by
Creating-Design
P.O. Box 1785
Columbus, IN 47202

Dedication

As you will read, my three sons, Eric, Chris, and Brian have been part of many of the adventures found in this book. Each have written a forward, appearing on the following pages. If asked about the proudest attribute of my life, without hesitation, I answer, Eric, Chris, and Brian!

Forwards
By

Brian Stark. born February 18, 1972 - Chris Stark, born June 29, 1968 - and Eric Stark, born June 15, 1966

(Left to Right) Chris, Brian, and Eric

Brian

My dad loves to tell stories. One could argue that when he left me, his youngest son, just seven years old, standing at a gas station in the middle of the night while he cruised down the road unaware for 500 miles, that it was all in the name of good story-telling later on. He claimed it was accidental and now, at 44 years old with 2 children and a wife of my own, I'm starting to believe him. Stories are a part of his soul. And he has spent the past 70+ years seeking them. Through hobbies as diverse as rating golf courses for the USGA/PGA to ham radio operator. He's participated in sports ranging from running marathons to sailing around Cuba to in-line skating, which he took up for the first time at 73.

What drives people to continue a search for something they've been seeking their whole lives? When does it stop? In my dad's case, it simply doesn't. Because what he's seeking isn't a map full of check listed tourist spots. He doesn't have a magnetic map on his Honda Goldwing motorcycle that he fills in with each 6,000-mile trip. My father is in search of stories, in whatever form they come. Whether it's from the guy who sold him a gallon of gas from his airplane hangar in the middle of Nevada or the soda fountain lady inside the hardware store, between aisles 12 and 13 in central Wisconsin. He loves it all. And because his goal is to accumulate as many stories as possible, he'll never run the risk of running out of them. Adventure hiker and author George Meegan said that the hardest thing to bear in life is never to have lived your dream and that the second hardest thing is to already have lived it. Fortunately for my father, George's words don't sound an alarm. He is not only living his dream right now, but he has been doing so for seven decades. How many of us can claim that? And what's even better is he's also found a way to ensure its longevity.

What transpires in these pages is an endless search for natural beauty, knowledge of history, and a connection to past friendships forged decades ago yet so familiar it's as if they can complete each other's sentences half a century later. Jim Stark is on a hunt. In search of something never quite attainable - he's not only quenching his thirst for knowledge and beauty, but also a lifestyle. To spend the day going through aerobic maneuvers on his motorcycle of banking, looking, leaning and reacting, all while feeling the sun on his face and wind on his skin only to be topped by a serene campsite with his folding chair, laptop, waterproof tent and a freeze dried dinner under the stars. Were it not for the comforts of home, which beckon after a few weeks on the road, a desire to share these stories with his supportive spouse and unmistakable yearning for a clean collar, we might not see much of my old man.

It's an interesting thing to grow up with a story-seeker for a father. It not only made for entertaining dinner conversation, but I found myself seeking stories in my own way at an early age as well.

But his motorcycle does point homeward from time to time. And when he does return he relives the whole ordeal over again, putting it down on paper, editing, re-editing and researching more parts of the story as-yet untold. It's our fortune that he suffers from this affliction. Each chapter of this book is another ingredient in a giant pot of delicious stew. Eat well, friends, and help yourself to a second serving. There's plenty to share.

***

Chris

Many times forwards are written by someone well known...a published author, celebrity, expert, etc. I can assure you I am none of those. However, I am a firsthand witness to most of the stories in this book, and a product of growing up in a home led by parents with a zest for adventure.

As Jim Stark's middle son, I vividly recall playing in our driveway with my friends, watching dad disappear down the street and return ten minutes later during his initial one-mile runs around the block. He even let us tag along on bikes or foot from time to time. However, it was not long before he figured out the "running thing" and would disappear down the same street not to return for three-hours as he knocked out twenty-mile marathon training runs! We decided staying in the driveway and playing kick-the-can was a better option for us than joining him on those!

Running was not all that common back then, and dad gained status as a local running celebrity as he logged mile after mile through town. It was a source of pride to have friends and grown-ups ask me about my "running-man" dad. Not surprisingly, I caught the running bug in grade school, and it became my sport throughout middle school, high school, college, and beyond.

Dad and I shared many hours together on long-runs while on vacations and training for running adventures. It is hard to imagine many fathers and sons having the same type of one-on-one time and connection that dad and I enjoyed through our mutual love of running. As we became older, we transitioned to hours on the golf course instead of the road, but we sure love to reminisce about our memorable running experiences.

As a collegiate football player, navy pilot, and successful businessman, I'm sure dad has always had great discipline in everything he did. However, as I was growing up, his dedication to running was the most visible demonstration of his discipline to me and my brothers. Dad's example of creating training plans, getting up before the sun to put in the miles, and meticulous documentation of his results, had an unspoken influence on the three of us beyond what we probably even recognize.

It has been a blast reading the running chapters in this book and re-living the stories. I would love to find dad's doctor friend who introduced him to the "cutting edge" concept of cardiovascular exercise back in the 70's, so I could thank him for what became a life changing, and lifelong activity for our entire family.

***

Eric

PILOT'S (READER'S) CHECKLIST

*items for conducting a safe (enjoyable) reading
of the following pages*

Checklists and flying go hand in hand. Whether the plastic laminated kind or some kind of memory device, a safe and enjoyable flight is ensured by the items on a checklist, either printed or virtual.

So it is with reading about these adventures, especially the chapters about flying! Your enjoyment of those pages—my father's adventures in the air (with the occasional "assistance" of his first-born son…er, me!)—will be enhanced by close attention to the following points.

Checklist Item #1

Receive the "I love flying" gene from your parent.

You can do this by admiring the airplane models and photos your parents displayed when you were young. In my case, Dad's den was a wonderland of elegant aircraft bristling with power. Whether the Navy aircraft he flew (see chapter three) or the detail of every scale model (including a small plastic circle glued to the nose of one plane to simulate the spinning prop disc...wow!!!), these images sowed the seeds for a lifelong dream.

Checklist Item #2

Feed your flying addiction by being schlepped across many county and state lines in pursuit of airplanes in the sky...whether scale models or life sized.

Back before the days of children strapped into car seats with fighter jet-style harnesses, my brothers and I tumbled down the interstate on our way to airshows, rallies and R/C parks in pursuit of all manner of flying craft. Whether Dad's own model aircraft (my memories are two-fold: first, how beautiful each plane looked when Dad was finished building it, and second, how ghastly his forefinger looked each time he would start the engine by spinning the prop with his digit...gross!). A childhood filled with visits to the EAA Airshow (now "Airventure" in Oshkosh, Wisconsin) can't help but amp up one's flying fever.

Checklist Item #3

Spread your own wings by doing some research.

It was ninth grade, and every one of us was going to have to write a real RESEARCH PAPER. This annual project was the stuff of legends in the school cafeteria. Once I hit ninth grade myself, "I thought I might quit school and join a cult.

Or would I? Actually, my fear of boredom led me to a different idea...what if I could come up with a research topic that was something fun? What if I could do an entire paper on FLYING?? Hours of library time would be spent looking at pictures of airplanes. Woo-hoo! And thus, my ninth grade research paper, examining the recent rash of accidents involving DC-10 wide-body passenger jets, was born.

I think this was a good plan, since I started bringing home library books and 3x5 cards all about flying. Dad started to notice, and soon he invited me to go flying with him (WIN! WIN!). (See chapter nineteen.)

Checklist Item #4

Have the good fortune to have a dad...er, "flight instructor"...who is extremely patient, and a LOT of fun.

Once Dad and I started flying together regularly, there must have been times he felt like he had opened a Pandora's box of inquisitive energy. I was devouring books about flying...how to do it, how the gauges worked, navigation. And I was constantly peppering him with questions. For a while, he might have enjoyed my keen interest in something he loved. But really, an 8-year-old with non-stop questions HAS to be a little much, right? Well, my Dad took it all in stride, patiently explaining, showing, teaching and demonstrating.

The times that were most fun, naturally, were the times spent in the air together. Learning maneuvers, sharing the joy of being airborne. But even then, there were times that I'm sure he must have wondered about the wisdom of stoking my enthusiasm.

One hot summery day we were out in the 172 and he was practicing some basic moves. I had never had any issues with airsickness before, but for some reason this day was different. It was like a movie in slow motion. Over a period of minutes, the nausea rose within me, the cold sweat, and the panic of knowing I was likely going to have to 'fess up if the flight didn't end soon. "Let's go over there and do steep turns above that bend in the river!" "Sure thing Dad, don't mind me while I study the stitching on my seat cushion." "Okay that was great! Let's head back to the airport and do some touch and goes!"

Unfortunately, I did toss some cookies during that flight. But Dad was a Prince, saying it wasn't a big deal, and ending the flight quickly. He made what could have been an awful day a lot better.

In the "fun" category, Dad was a winner. After numerous hours of flight training, Dad offered me the chance to fly left seat (the pilot's traditional seat). WOW!! The feeling of achievement, finding the ignition key right where it belonged in my left hand, the throttle in my right, keying the mike, … I WAS the REAL PILOT…what a RUSH!

Another fun flying day: doing a couple turns in the pattern with my door off. It's a long story, but I don't think I ever cinched my seatbelt as tight as I did that day. With no door at my right elbow, I didn't want to take a tumble. But it was great! Even the light rain shower didn't do much to dampen our spirits that flight.

Checklist Item #5

The love of flying comes full circle

The pressures of graduate degrees and career had pulled me away from flying for years. But one day, feeling especially sorry for myself since I was swamped with work, I had had enough. I said "If a guy with five part-time jobs can't go flying, then no one can." That very day, I toured the local flight schools and a week later had my first lesson. Thanks to my Dad, I soloed after only about 7 hours of dual, and about 9 months later, I had my pilot's license!

Though I was a bundle of nerves, it was a special joy to take my Dad flying the first time. Holy cow, was I crazy?! Why would I invite a former Naval aviator to fly with me??!! These guys nail airspeeds like a Swiss watch…my meandering technique would be no match for that. I was really nervous. But, after flying Dad and stepmom Michele to lunch, we had so much fun, I soon relaxed and felt at ease.

That flight, and others we have shared, was a milestone I could never have imagined years ago, looking at Dad's airplane pictures and models. Each shared experience is a gift, as you'll read (see chapter twenty-six). Dad was always the path to flying, the guy who opened the door and showed me the way. And now that I've been flying for twenty years myself, flying is and always will be a connection to my Dad. His lessons about safety, the priceless adventures that spanned a continent, and the special bond between an airborne father and son, these are the gifts I find reflected in those flying chapters, and for which I give thanks every day. Enjoy the journey, and have a safe flight!

True Adventure - Chapters

Chapter One	"Call Me Lucky"	Page 1
Chapter Two	"So it Begins"	Page 6
Chapter Three	"Navy Times"	Page 12
Chapter Four	"First Steps"	Page 23
Chapter Five	"Where's Brian"	Page 32
Chapter Six	"Setting Sail"	Page 37
Chapter Seven	"ABATE"	Page 46
Chapter Eight	"Motorcycling Buddy"..........	Page 51
Chapter Nine	"Flight Training Excitement"	Page 61
Chapter Ten	"Rosey"	Page 68
Chapter Eleven	"Yellow Blood"	Page 81
Chapter Twelve	"IJDGABTT"	Page 87
Chapter Thirteen	"Sedona"	Page 98
Chapter Fourteen	"Are We Having any Fun Yet"	Page 102
Chapter Fifteen	"Humor in Uniform"	Page 109

Chapter Sixteen	"Heavy Mileage"	....	Page 119
Chapter Seventeen	"Westward Ho"	....	Page 124
Chapter Eighteen	"A Funny Thing Happened"	...	Page 133
Chapter Nineteen	"Civilian Aviation"	....	Page 146
Chapter Twenty	"Chasing the Sun"	....	Page 154
Chapter Twenty-One	"Obsession"	....	Page 163
Chapter Twenty-Two	"Adventure on the High Seas"	...	Page 169
Chapter Twenty-Three	"Canyon Conversations and Bambi"		Page 189
Chapter Twenty-Four	"Return to Paradise"	....	Page 198
Chapter Twenty-Five	"Running on Empty"	....	Page 214
Chapter Twenty-Six	"Decathlon Aerobatics"	....	Page 219
Chapter Twenty-Seven	"On the Road Again"	....	Page 226
Chapter Twenty-Eight	"Pamplin Historical Park"	....	Page 233
Chapter Twenty-Nine	"First and Last"	....	Page 239
Chapter Thirty	"Hang Gliders"	....	Page 254
Chapter Thirty-One	"Ike's Place"	....	Page 260

Chapter Thirty-Two   "Glacier National Park" ......   Page 267

Chapter Thirty-Three   "Travel Highlights" ....   Page 277

Chapter Thirty-Four   "Sailing the Greek Isles" ....   Page 288

Chapter Thirty-Five   "Doing the Big Apple" ....   Page 317

Chapter Thirty-Six   "Magic Carpet" ....   Page 325

Chapter Thirty-Seven   "Wyoming National Parks" ...   Page 344

Chapter Thirty-Eight   "Pony Express Trail" ....   Page 351

Chapter Thirty-Nine   "The Finish" ....   Page 359

Chapter One

Call Me Lucky

It was cool that October morning as I rode my Yamaha motorcycle from my home in Columbus, Indiana, to our Bloomington office forty-five miles away. Dawn had not broken and I shivered, hunched over the handlebars anxious to arrive before the 7 a.m. meeting. As I sped along at seventy miles per hour, the deer's head suddenly appeared in the beam of my motorcycle's headlamp, three feet away. I remember its eyeball, large and red-rimmed, reflected in the light. The impact was instantaneous – no time to close the throttle, let alone apply the brakes.

I don't recall the moment of collision but assume the deer must have hit the motorcycle broadside, rather than me hitting it, because I wasn't thrown forward. Instead, the bike was knocked out from under me. My first awareness of the accident was sliding, not tumbling, down the highway. I saw the motorcycle scraping along beside me, sparks flying, and thought, "This isn't so bad."

I was wearing leather chaps and motorcycle boots that chilly morning, having put my dress shoes and suit jacket in the saddlebag. A motorcycle jacket covered my white shirt and tie. I had three different helmets to choose from, and although my least favorite is the full-face helmet protecting my jaw, I had worn it because of the morning's brisk temperatures.

When I came to a stop, I sat up, shook my head and arms, and announced to whoever it is guarding hapless souls, "I'm okay." Then I noticed cool breezes. I didn't have any pants on. The fifty-yard skid down the highway had stripped off the leather chaps and tattered my dress slacks. The slide had filed off the toe of one boot. The sleeves of the motorcycle jacket were in ribbons, like those of my dress shirt underneath. The jaw portion of the helmet was broken, one side completely separated, and the other cracked. Apparently, I smacked my head when going down. If I had worn any other helmet that morning, I would have had a rather damaged face. Other than road rash and abrasions, I was in remarkably good condition.

Call me Lucky. I always felt that ought to be the title of my autobiography. I have been one lucky guy throughout my life. Being able to walk around following the wreck I'd just experienced was miraculous. Motorcycles are dangerous. I'm aware of that threat. Something as minor as a fender-bender in a car can cripple a motorcycle rider. More than 2,200 deaths occur nationally every year in motorcycle accidents. I have talked more people *out* of riding these machines than I have tempted *into* joining the ranks. Michael Bamberg is an example.

Michael joined our inline skating club one morning in Florida. His wife, Lois, is a writer friend of mine. Mike made quite an entrance. Our group was getting ready to skate, tightening helmet straps and adjusting knee pads when Mike came rolling up and fell flat on his butt. Embarrassed, he scrambled to get up but tumbled over once again. We all looked at one another and wondered, "Is this guy going to skate with us?" Yes, he was. What a disaster. He plummeted into bushes, crashed into fences, slipped off hard surfaces and caught the grass, which sent him sprawling head over wheels. How he made it back to our starting point that morning without a broken neck, I have no idea.

I had ridden my motorcycle over to our gathering place. When Mike saw my parked machine, he said, "Man, I got to get me one of those." He had to be kidding. He'd be killed before he got it home from the dealership!

I said, "Mike, motorcycles are one thing you should avoid like the plague. Just one stumble like you had today could kill you. You think those knee pads and elbow pads will protect you? I've seen helmets torn in half after a motorcycle accident. I don't know of a serious motorcycle rider who hasn't had at least one accident. It's almost a certainty."

"Wow, I didn't realize," Mike said. "Maybe I better reconsider. I rode a bicycle at one time but gave it up after I broke a tooth and had thirteen stitches. I just thought motorcycles would be easier, not pedaling and all."

Whenever I see Lois, the first thing she says, is, "Thank you, thank you!"

The leather I wore that October morning literally saved my skin, although I had some minor wounds. Various points of contact with the concrete by my body — calves, hips, and forearms — were raw from scrapes after the leather had been ground away. Both wrists where the gloves met the sleeves of my jacket had weeping abrasions. My right knee had been dragged along the road and had a one-inch laceration that would require a few stitches. But other than that, I walked around, marveling I could move without pain. What luck! I pulled the motorcycle onto its wheels and started the engine. Even with its bent handlebars and several dangling parts, I contemplated riding it home.

A car stopped moments after the crash. The driver must have had a cell phone, because minutes later the police arrived. The policeman wanted to call an ambulance. "Oh, no," I insisted, "I'm fine." When I suggested I might even ride the bike home, he said, "No way. You may be in shock and not know it. You could get a mile from here and pass out. And as for riding that wreck, I couldn't allow it. Please let me call an ambulance. If nothing more, just to check you over. If you're okay, fine, we'll send it on its way."

"Okay," I agreed.

Then the policeman asked, "Do you want the deer?"

"Huh?"

"Yeah, the deer you hit, fifty yards up the road there. Where you collided. The dead deer. I have a long list of people wanting road kill, but drivers get first priority."

"Geez, no. I don't want it."

So the cop called the ambulance and then called somebody who picks up the road kill to take it to the top name on his list. The first to arrive was a little wrecker with a hoist to pick up the deer. It beat the ambulance, with its screaming siren and flashing lights, by ten minutes.

The two ambulance EMTs wrapped my abrasions, bandaged the cut on my knee, and then checked my heart and blood pressure, 120 over 80. All appeared normal. They couldn't believe I had just been in an accident. They, like the policeman, nevertheless urged me to let them take me to the Columbus hospital to be checked further. "And besides," they said, "you need to get that knee stitched up."

Standing there in rags for pants, I had an idea. "If you take me to the hospital, do I have to go in?"

"Well, no. But what will you do?"

"I need some clothes. Let me call my Columbus office. I'll have our delivery driver meet me at the hospital and take me home. Once I'm cleaned up and put some pants on, I promise to drive to the hospital for stitches and let them do whatever checking they need to do."

It was agreed. I bid adieu with thanks to the policeman, who by this time had called a body shop to haul away my bent — and later judged totaled — motorcycle to a garage. After my visit to the hospital and a few stitches, I went to the Columbus office. By this time, the entire company, it seemed, had learned about the accident. "Deer Slayer" became my new handle.

Now here's the rest of the story.

The primary point of contact as I slid those fifty yards down the highway was my left butt cheek. That's where I carry my wallet. The billfold had been given to me by the Harvey Hubbell Company, one of the manufacturers my company represented. Hubbell handed out wallets to its distributor managers associated with a promotion being launched.

The billfold became totally destroyed in the slide. Not only was it pulp, but my driver's license, Social Security card, and other contents had been ground into scrap. I don't carry money in my wallet, but Harvey Hubbell, as is the custom, had included a one-dollar bill, and I just left it in the billfold. That bill became just-barely-legal tender.

I sent the remains of the wallet, along with the story of its destruction, to Hubbell's vice-president of marketing. The final line of my letter said, "Thank you, Harvey Hubbell, you saved my ass!"

A few days later, I received a small box containing two billfolds. The note said, "Not only does Harvey Hubbell cover all bases, but in this case, both cheeks."

# # #

Chapter Two

So It Begins

This book about adventure is not about climbing Mount Everest, blasting off into space in a rocket, or fighting crocodiles in the Amazon. Hardly. The stories I tell are adventures of an ordinary man, one fortunate enough to have been a military pilot, traveled long distances by motorcycle, sailed on both coasts, and run more miles than the circumference of the Earth. Pursuing those avocations created more than a few interesting, humorous, and sometimes frightening incidents, like the time Soviet fighters attacked me during the Cuban Missile Crisis, or sailing onto a reef in the Caribbean Sea, or running the equivalent of six back-to-back marathons. Those episodes are the adventures I will share in this memoir.

Another of my hobbies is journaling, keeping records of my activities. I'm a nut about it! My records go back to the early 1960s. A random perusal of those log books shows that on Sunday, January 7, 1979, I ran 6.1 miles. It snowed that day, and there were six inches of snow on the road, so footing was terrible. The run took fifty-five minutes and fifteen seconds.

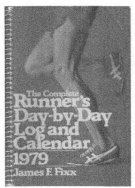

Another journal reveals that on April 3, 1985, in Florida, I scored an eighty-one on a golf course. I hit eight fairways, was on the green in regulation ten times, and had thirty-three putts. The document for April 4, 1963, shows I flew a Navy airplane for 1.3 hours and made eight touch-and-go landings. Those accounts are in my journals. Pretty compulsive, huh? I just enjoy writing things down.

When I travel by motorcycle, I share my sights and experiences by emailing daily reports to sixty-six VIPs, my Virtual Invisible Passengers. The reporting, sharing my stories, is an important part of the trips. During the day, as I witness breathtaking landscapes, I contemplate how I will describe them later via email. One or two hours at the campsite each night, with my computer in lap, is spent documenting my experiences. If my camera or laptop breaks, as has happened on occasion, my role as a reporter is threatened, and repairing those tools becomes an urgent priority.

I have documented my adventures in scores of journals over the years. A four-drawer file cabinet overflows with the booklets. It's from those records I pull my stories. This won't be a chronological account; I intend to jump around, devoting each chapter to different episodes.

Let me start by telling you a bit about myself. I was born in Connecticut in 1938. Whoa! I'm an old guy! Yeah, but still doing okay, as my stories will illustrate. I had a rather exciting early beginning. The *Bridgeport Post* in 1939 reported that, as a fifteen-month-old toddler, I was run over by a car. I had chased a dog, unaware of a man backing out of his driveway. The driver stopped when he heard my mother scream, "My child!" Looking under his car, he saw my hand pinned under a wheel. After driving forward, he took me to the Bridgeport hospital, where, per the news article, I was treated for injuries to my head, ribs, and hand but was reported to be in "fair" condition.

My first flying lesson.

At four years old, I responded to the challenge of my playmates to ride my new tricycle down a big hill and collided with a truck just entering the intersection at the bottom of the hill. The truck driver stopped when he spotted me careening toward him. The tricycle struck just forward of the rear wheels, and I went underneath. Pulling me from under the truck, I was identified by my playmates and the driver carried me up the hill to my home. My friends ran ahead to announce to my mother, "Jimmy's been hit by a truck! Jimmy's been hit by a truck!"

In pictures of my growing-up years, I always seem to have a bandage somewhere on my head or limbs. I was a rough-and-tumble youngster. When I was not yet two years old, my dad used to balance me on his hand, then throw me high in the air.

In 1937 my dad became an executive trainee of the General Electric Company. We lived in Bridgeport, Connecticut, GE's world headquarters. During my first five years, we moved seven times, because of new GE assignments to different homes in Connecticut, Virginia, and Maryland. That was just the start of my nomadic life. The other GE transfers had us living in Washington, D.C., Indiana, Missouri, and Nebraska. To date I've had thirty-eight permanent changes of address. Some of those were thanks to my service as a pilot in the United States Navy.

When growing up, it seemed I was always testing my nerve and daring. Could I jump off a roof, or stand on the seat of my bike and coast down the hill, or leap off the twenty-foot swim tower into the water? My mother was a registered nurse, and you would think she would be used to the sight of blood, but every time I came home with new wounds, she panicked. I couldn't understand it. "Mom," I would say, "before I pull up my pant leg and show you, it's really not that bad." Then she would notice the blood pooling on the floor and start screaming.

I seemed to be drawn to the activities of hard knocks. Perhaps that's what led to a scholarship as a college football player.

So, with this history of bumps and bruises, what would be the worst possible activity I might pursue? How about riding a motorcycle?

I really didn't get into motorcycling until my forties. And like so many things in my life, it happened as a fluke. I had been reading a book titled *How to Get Anything You Want* and decided to try its methods. First, it said, announce your goal.

But wait! First, what did I want? I had been reasonably successful in life if measured by job, money, home, and family. I really didn't crave a mega-million dollars, although according to the book, that was fair game. Let's not be greedy, I thought. I had never ridden a motorcycle, and it looked like fun. Okay, world, *"I WANT A MOTORCYCLE!"* Incidentally, that announcement did not include telling my wife. She wouldn't approve.As vice-president of an electrical supply company with branches in three states, I had a primary office in Columbus, Indiana. But I was responsible for eight other branches, one of them in Bloomington, Indiana. I mentioned my motorcycle interest at a meeting one morning in the Bloomington office. It wasn't an hour later that our delivery driver in that location said, "I hear you're looking for a motorcycle. I drove by a nice one for sale by the apartments over on Winslow Road. Check it out."

I did check it out, a 500cc Honda, and the sign said, "For Sale $600.00." I left a note. "This is a nice-looking motorcycle. If you have no other offers, I will pay $300. My phone number is 812-555-1234."

I got a call the next day, and the motorcycle was mine.

Since I had never ridden a motorcycle, I had a Bloomington employee bring it over to the Columbus office, where I hid it in our warehouse. He rode it over with his wife, and I repaid them by flying them back to Bloomington in my airplane, doing some sightseeing en route.

Late one afternoon, after all the Columbus employees left for the day, I wheeled the Honda out of the warehouse onto our parking lot. After starting the machine, I slowly rode across the lot and gingerly made the turn at the far end. I repeated that crossing several times until I felt comfortable. Okay, I was ready for the big test. I took the motorcycle onto the country road adjacent to the office and headed off. Time to find out what this baby could do. I rolled the throttle open and held on. Tears were streaming down my face, and my cheeks were inflated by the wind. I was breathless with exhilaration. Sneaking a look at the speedometer, I saw I was screaming along at twenty miles per hour!

Thus began my love affair with motorcycles.

A few weeks later, having enjoyed secretive after-hour rides on my exciting new toy, my wife and I were going somewhere in the car. Unexpectedly, she asked, "Do you have a motorcycle?"

After a pause, I replied, "Oh sure, I have a motorcycle. Yeah, right! Me? Yeah, I have a motorcycle."

She said, "Well, I didn't think so, but Roseann Watson said she thought she saw you riding one."

I had been taking my youngest son, Brian, on rides, so it wasn't long before he slipped up, and the truth became known. As expected, my wife was not happy.

"Jim, you're going to kill yourself. I hope you know that!"

She was wrong. I didn't kill myself.

But I did have some close calls and many years of fascinating experiences.

# # #

Chapter Three

Navy Times

I equate the skills required to fly an airplane to those of playing a musical instrument. A six-year-old child squeaking out "Twinkle, Twinkle, Little Star" in a Suzuki violin ensemble is quite different from the first violinist in a professional symphony orchestra, but they both play. In a similar fashion, almost anyone can be taught to maneuver a Piper Cub off the ground, circle the airport, and land safely back on the ground. But navigating across hundreds of miles of water and making a night landing in a complex jet fighter onto a pitching aircraft carrier deck would test the skill of the most experienced professional pilot.

I use a Navy carrier pilot as an example of peak proficiency in airmanship because of a smug prejudice that elevates the "boys in blue" to the very top of the heap. The best airline captains come out of the military, and many of them are former Navy. My brothers in the other branches of the military who fly will challenge that of course, but you can understand my prejudice because I wore Navy wings.

New Ensign USN

The Navy training I received, including the abovementioned aircraft carrier landings, enabled me to fly across oceans, challenge the extremes of Mother Nature, and bring my aircraft and crew safely home during two thousand hours as pilot-in-command.

The Navy gave me a gift. Not just that of flying an airplane, but a gift of confidence, a positive attitude, and a "can do" spirit. That wasn't always the case. I struggled academically as a youngster.

My father's every-year GE transfers to a new state became a challenge for me to get on the same page with my classmates in a new school. Perhaps I should have repeated a grade somewhere along the line, but I kept getting promoted. After I graduated from grade school with marginal grades, my parents hired a tutor to get me qualified and enrolled in the challenging curriculum of the Fairfield College Preparatory School, a Catholic prep school with a stringent Jesuit curriculum. Slapped into shape by the demands of the priests at Prep, I did okay, even earning academic honors during a few semesters, but I didn't do much maturing. My hard-nosed athletic aggressiveness led to all-district honors on the football team, and my toss of the discus during track season set a school record. I timed that record-setting achievement perfectly, as University of Pennsylvania football recruiters visited that day, eyeballing me for their football program. The effect of all this confirmed in my mind my status as a hot shit and that I could look forward to a college life of beer and pretty girls.

The Wharton School of Business at the University of Pennsylvania was not impressed by me. Overwhelmed by freshman football, fraternity activities, and demanding studies, by the time I settled down academically and opened the books, I had dug myself into a hole of failing grades.

"Mr. Stark, you're out of here!"

Okay, loser, what do you do now? I asked myself. Fortunately, my dad's alma mater, North Central College (NCC) in Naperville, Illinois, expressed interest. With no grades to transfer, I enrolled the following fall. Years later I learned that my dad's brother, also an NCC grad, had approached the coach of North Central's football team, promoting my availability and hyping my football prowess.

North Central College 1962

I got along sufficiently at North Central College as a "C" student but for that reason had a rather "average" opinion of myself intellectually. The Navy changed that.

I entered the Navy's flight program almost by accident. As a junior in college, I had no plans after graduation but knew I wanted no part of that Vietnam mess heating up.

Two Navy recruiters, resplendent in their blue uniforms with gold braid around each sleeve and gold wings above their breast pockets, walked into our student union one afternoon and set up their table and posters. Students avoided them like a bad smell. A career in the military had all the appeal of itinerant farming for most students. Sitting in a booth drinking my chocolate milkshake, I felt sorry for these guys. My dad had skippered a PT boat in World War II, and I had always thought highly of the Navy. I went over to visit with them.

I remember little about that conversation other than that they suggested I drive over to Naval Air Station Glenview, only an hour from campus, and take a ride in a Navy airplane. Sounded like fun, so one Saturday I hopped into my '57 VW

Beetle and drove to NAS Glenview.

The Marine at the gate directed me to a building where a number of other college-age guys had gathered. I can't believe how naive I must have been. Although I didn't know it at the time, those other young men had been flown in from universities all over the country. I took some tests, got a physical, and was told to wait for an interview. Sure a lot of bullshit just to get a ride in an airplane, I thought.

Finally, ushered into an office with two senior naval officers seated behind a table — I supposed they must be admirals — one of the officers leaned back in his chair with a challenging scowl and said, "Tell me, son, why do you want to be a Navy pilot?"

"Navy pilot? Geez, I only came up here to get a ride in an airplane!"

The demeanor of my two inquisitors immediately changed. They suddenly became aggressive salesmen promoting, they said, the most prestigious life the military had to offer. "See the world, fly with the eagles, be part of a proud military brotherhood!"

It must have worked. I didn't get an airplane ride, but I have an eight-by-ten glossy photograph showing me with my right hand raised in commitment.

Now do I get my airplane ride?

I did well in the Navy. The initial four months of officers training in Pensacola, Florida, included physical training, military indoctrination, and academics. The academics involved aviation-related subjects such as aerodynamics, but it also included calculus and military history courses. We took tests frequently in academics as well as PT and military training, and I earned a badge to wear on my uniform indicating excellence in all three of those activities.

The following fourteen months of pilot training also included ground school, and I found myself excelling in those schools as well. Perhaps I wasn't as average as I believed.

Initially, my flight training didn't go well. My fellow student pilots all talked with excitement about their first flights, and how cool everything was. I dreaded every minute I spent with this asshole Vitacco. I thought I was going to wash out and really didn't care.

Flight instructor Ensign Vincent Vitacco was a pissed-off, newly commissioned officer, who, after getting his wings, instead of being sent to the fleet and becoming a war hero — in his mind — got orders back to the training command as an instructor. Vitacco, no older than I, was a screamer. His teaching technique was to browbeat and verbally assault his students. On our first flight together, while we taxied out for takeoff, he described the use of the toe brakes on the rudder pedals. Sitting in the cockpit in front of him, I looked down and watched the pedals as he described their use. From the back cockpit I heard, "What the hell are you doing, reading something? Pay attention, Goddamnit!"

"Sir, I was only watching the rudder pedals."

No reply.

During that flight he demonstrated S-turns. As we turned in one direction, I instinctively leaned in the opposite direction.

"Oh, afraid you're going to fall out, are you?" And with that he rolled the airplane upside down.

I must admit it was upsetting, in more ways than one.

In addition to putting up with him, I had trouble with ear blocks. We weren't flying high, but after the flight, no matter how hard I pinched my nose and puffed my cheeks, I couldn't clear my ears. I finally had to see the flight surgeon.

The doc said ear blocks could result in serious issues. If not cleared, blood would be drawn into the eardrum and cause an infection. The surgeon's solution was to shove an air hose up my nose and have me gulp a glass of water. When I swallowed, the doc hit the trigger on the air, and I thought the top of my head was going to blow off. It cleared my ears, all right! I can't imagine that's the same treatment used today, but it worked.

Afterward, a cowardly thought occurred to me. If I got a medical discharge from flying because of my ear problems, what disgrace could there be in that? No more putting up with Ensign SOB! Friends and family would certainly understand, wouldn't they? Of course, I wouldn't be sent home; I had enlisted in the Navy and would be put on a ship somewhere. But how bad could that be?

I was about to fly my first solo, so I thought I would stick it out a bit longer. I was actually getting good flight grades from Vitacco. Compared to my friends', my grades were all above average. If I was doing well and still catching hell, I wondered, what about his below-average students? He must be killing those guys!

Student pilots ready for their first solo were given a check ride by a different instructor. During the flight he would observe your basic maneuvers and evaluate your skills during touch-and-go landings. If he determined you were qualified, you would make a full-stop landing, the instructor would get out, and you would make your first solo take-off and landing. My buddies and I all agreed: the smart move with the instructor was to demonstrate how safe we were.

When at last the big day arrived, I met Lieutenant Shepard, the check-ride instructor, at the flight line, and I couldn't believe how nice he was. He called me Jim right off and briefed me on the flight's procedures. "No need to be nervous," he said. "We'll be doing the same maneuvers you've been practicing over and over for the last few weeks. You'll do fine. Let's go flying!"

Wow, what I would give for an instructor like him!

We took off, I did the drills, and then the time arrived for the touch-and-go landings. Here's where I would demonstrate how safe I was. As I turned onto final for the first approach, I said, "Sir, my airspeed is a tad high. Think I'll take it around for another approach."

"Okay."

On the second attempt, I said, "Sir, my altitude is a bit low. Think I'll take it around for another try."

On the next turn to final, Shepard interrupted. "Jim, your approaches are fine. What say you get this damn bird on the deck, so I can get out, and you can qualify?"

"Yes, sir."

Lieutenant Shepard got out, and I flew off on my own. Frankly, it didn't feel much different from any other flight. Sitting in the front seat without the instructor behind me certainly didn't look any different from a normal practice flight. I could tell the plane was a little lighter when landing, but the solo was no big deal.

I felt it was a bigger deal on my next flight, when for the first time I'd be doing the entire flight alone, from preflight inspection, flight maneuvers, practice landings, to the return to home base.

Doing the preflight inspection that day, I looked that airplane over like I was going to buy it. I checked every rivet. Once at altitude, I did clearing turns, looking for other aircraft like I expected Japanese Zeros to drop out of the sky above me. My practice landings were flawless. When returning to home base, I keyed the microphone, used my deepest, most professional airline pilot's voice, and requested landing clearance. Yes, by God, I was an aviator!

That flight was a turning point. I didn't fly with Vitacco much anymore. Half my flights were solo, and the ear problem went away, never to return. And I was introduced to aerobatics: loops, rolls, spins, Immelmann turns, Cuban Eights — damn, I loved that stuff. I thought hanging upside down in the straps, with change falling out of my flight-suit pockets was about as good as it got.

As I moved on in the training command and later into larger, multi-engine airplanes, aerobatics were no longer possible. The big planes we flew would not have been able to handle the forces of a roll or a loop. But our first trainers, the Beech T-34 and the North American T-28, were built to endure those gravitational (G) forces.

When pulling out of the bottom of a loop, it's likely you will be pulling three Gs, a force three times normal. A two hundred-pound pilot feels like he weighs six hundred pounds. Three Gs is quite normal in aerobatics, and up to six Gs can be tolerated for a few seconds.

When you pull positive Gs, as in at the bottom of a loop, all the blood in your upper body is pulled toward your feet. If those forces continue too long or are extreme, you black out. Jet pilots wear G-suits that cinch tightly around their torsos to restrict the flow of blood and limit the possibility of blackouts.

Negative Gs occur in the opposite direction, as when flying upside down and blood is forced into your head. This can cause a "redout" and can be harmful to brain, eyes, and ears.

The T-28 Trojan we flew had a 1,600-horsepower engine. You could climb straight up for a short time before running out of airspeed. The T-28 had one danger, however, and that was a tendency to enter a flat spin.

A flat spin begins with a stall. A stall occurs with the interruption of airflow over the tops of the wings, as happens when becoming too slow. In that situation, the aircraft loses lift, the plane stalls, and the nose of the plane drops. If it drops straight down, the airflow will increase once again, and the plane will recover, with almost no corrective action by the pilot. This assumes there is enough altitude for that recovery to occur. Unfortunately, a lot of stalls occur during landings close to the ground. Low *and* slow is a killer.

If at altitude and the plane stalls, and if it falls off on one wing, either by accident or by the pilot's intention, it will enter a nose-down spin. The spin occurs because as the plane rotates the fuselage blocks the flow of air over the wings, and if no corrective action is taken, the plane will spin continuously into the ground. The corrective maneuver is to push the stick forward for a vertical dive and stomp on the opposite rudder to stop the spin. We learned spin recoveries and practiced them often.

When the T-28 would stall, because of its unique weight and balance, it had a tendency to go into a flat spin and rotate horizontally. When that happened, wind over the rudder and elevators became blocked, and there was nothing the pilot could do except bail out.

I was scheduled to fly a solo in the T-28, but we had a week of rain and low ceilings at Whiting Field, and no one could get into the air. At last the weather cleared, and the loudspeaker announced, "Launch all solos!" The sky became full of dozens of student pilots, all looking for clear airspace to practice their maneuvers. I took off and flew to a far corner of the flying area to get as far away from other airplanes as I could to practice aerobatics.

I started with the loop. The procedure was to first make ninety-degree clearing turns both right and left, looking for other airplanes. Every time I did that, I saw another airplane flying too close. I began getting impatient. Finally, perhaps not turning a full ninety degrees, I decided the sky looked clear enough. I shoved the stick forward to pick up the needed airspeed, enough to pull the plane into a climb, and high enough to pull it over on its back to complete the loop.

When I reached the required airspeed in the dive, I hastily pulled the nose upward. This is a positive-G maneuver, and we were supposed to keep that lifting climb to under three Gs. A G-meter on the instrument panel indicated the G-forces.

In my haste to avoid other airplanes and do the maneuver quickly, I pulled the stick back too far. I could feel my head becoming light, and when I looked at the G-meter, it indicated six Gs. Oops, way too much, so I tried to relieve the pressure on the stick, but not soon enough. My field of vision was shrinking, as if I were looking at a small TV screen, a physical sign of impending unconsciousness. Then everything went black.

I don't know how long I was out, probably only a few seconds. When I came to, I shook my head, trying to figure out what was happening. I looked out the canopy at a confusing scene. Then I realized the plane was pointing straight up. A quick glance at the airspeed indicator was even more puzzling; it pointed to zero. I was hanging on the prop. The tail slide didn't last long, and slowly the nose fell through, only it appeared the plane was twisting onto its back.

I kicked the rudder pedals first one way, then another, and slapped the stick in all directions. But nothing happened, because the plane was too slow. Somehow the aircraft rolled right side up and started to rotate, but in a flat configuration. Oh, no! The dreaded flat spin! The Navy gets really upset when you walk into the operations hangar, dragging your parachute, and report your airplane is buried in some farmer's field.

Everything happened in slow motion. I remember holding the stick full forward and lunging forward against my shoulder harness as if my body weight would help put the plane into a dive from which I could recover. Still I rotated, slowly, mostly horizontally. I reviewed my bail-out procedures. Where was that emergency canopy release handle?

As I was trying to sort out my dilemma and tighten my parachute harness, the plane, on its own, dropped into a standard nose-down spin. Hallelujah! I went through the procedures, recovered, leveled off, and began breathing normally. No, the seat wasn't wet; I checked.

I said to myself, "You know, I think that's enough for today. Think I'll go back to base and land."

Turning in my parachute, I ran into a friend. "How'd it go?" he asked.

"Oh, okay. Few stalls, spins. You know, routine stuff."

T-28 "Trojan" Whiting Field

\# \# \#

Chapter Four

First Steps

It was one of those insignificant moments that happened with little fanfare yet influence your life in momentous ways. At a party in 1974, a doctor friend got my attention when he said, "Jim, I discovered this great exercise program called 'Aerobics' by Ken Cooper. You'd love it. 'Aerobics' means requiring oxygen to meet energy demands during exercise." He was right, I did love it! The program resulted in my obsession over the years to run more miles than the circumference of the Earth.

First thing Monday morning after the party conversation, I bought Cooper's book *Aerobics*. The initial step, according to the program, was to determine one's fitness condition. A simple test suggested running or walking as fast as possible for twelve minutes. How far you traveled in that length of time determined your fitness level. I took the test and covered just short of a mile. Not bad for someone who didn't exercise regularly. The program then recommended, based on age, health, and other factors, target aerobic exercises measured in point values needed to achieve and maintain physical conditioning. A variety of exercises could be engaged in for accumulating the points: running, walking, bicycling, swimming, playing singles tennis, walking a golf course, and others.

I love keeping records and admit I'm a fanatic about it. For instance, there's my statistics on each round of golf I've played over the last twenty years, recording fairways hit with my drive, greens reached in regulation, putts, and other details. So, recording aerobic points was my hot button. I still have those running records dating back to 1974. I did a variety of exercises initially, jotting down each day's points in my logbook. However, I found running to be the most efficient. A one-and-a-half-mile jog equaled walking two hours on a golf course.

In the beginning, I ran three miles a day. As I became more conditioned, I loved the feeling of speed and of being athletic, and the healthy fatigue of exercised muscles. Being able to run those miles faster each time was exciting and earned more points! My pace went from ten minutes per mile, to nine, and eventually to eight.

Here's a confession; I was a smoker. First had been cigarettes in my twenties, but during this period, my late thirties, I smoked a pipe. I thought it made me look very distinguished, and people said they liked the aroma. But I inhaled the smoke, so the pipe wasn't any better, was perhaps even worse, than cigarettes. As an experiment, I decided to not smoke for one week, to see if it enabled me to run faster. The results were astounding; I improved my time per mile by over 30 seconds! I stopped smoking cold turkey and never lit up a pipe or cigarette again. That was over forty years ago.

Fat, dumb, and happy, 1973

My improved running performance made me think I had a special ability for that activity. I knew that as a college football player I wasn't considered fast in short-distance time trials, but maybe long-distance running was my thing. I remember the day I ran ten miles and told my doctor friend that in one day I met all my aerobic points for the week. The neighbors talked about my daily runs around our streets. "There goes that damn fool runner. What is he up to, anyway?" That's how unusual jogging seemed at that time.

My first race experience was a ten-kilometer race (6.2 miles) held in my town of Columbus, Indiana. I hoped to run at the eight-minute-per-mile pace I had achieved during my three-mile training runs, but how would I judge that pace in the race? Sports watches with their stopwatch and split-time features had only just made their appearance in sporting goods shops. Nike had recently introduced its waffle-soled running shoes, which were all the rage. I still ran in my Adidas tennis shoes.

Before the race event, I spotted Ray Sears, a seventy-year-old runner from Shelbyville, Indiana, who had been featured in *Runner's World* magazine for his senior-age running accomplishments. I slipped up close to Ray and overheard him say he ran at an eight-minute pace. Perfect, I'd simply follow Ray.

How exciting the day was. My running number had been carefully pinned to my event T-shirt; a sweatband coolly adorned my forehead. Friends and spouses gathered outside the ropes, giving their runners last-minute encouragement. My stomach fluttered with the familiar nervous jitters I remembered from athletic events years ago in high school and college.

With Ray in front of me as my rabbit, I was ready. When the gun went off, we sprinted ahead under the pennant-draped starting line. Within yards I found myself gasping for breath. Must be all the excitement, I decided. At the first mile marker, my lungs burned, and I panted like a hound dog chasing a Buick. I had started way too fast. Breathlessly waving farewell to Ray, whom I noticed casually chatted with fellow competitors as he cruised down the race course, I slowed, hopefully to catch my breath and to be able to finish. The last two miles were painful. My side ached, my legs throbbed, and my stomach felt like it might erupt. I saw Ray after the race by the Gatorade and banana table and asked, "Ray, I understood you to say you ran an eight-minute-per-mile pace?"

"Yes, right," he said, "for a twenty-six-mile marathon. For these shorter runs I let it out a bit!"

A mini-marathon in Indianapolis — 13.1 miles — became my next race. What? I run *one* six-mile race and think I'm ready for a half-marathon? A neighbor suggested it, but also my increased performance in training convinced me I could do it; every day I became faster and stronger. I ran thirty miles each week and had lost twenty pounds. I looked lean and mean and felt terrific.

Race ready, 1978

I did okay in Indy. Nearly a thousand runners ran the race. Some came from distant states and included world-class athletes. I enjoyed being part of such a spectacular event. The final two and half miles of the Mini finished on the Indianapolis 500 race track. When I entered the brickyard track, I approached the first turn wall, slapped it with my open hand, and spun around three times. Not many folks can say they raced at Indy, hit the wall, and spun out.

Shortly thereafter, on my fortieth birthday, October 5, 1978, I decided to run twenty miles, from the west entrance of

Brown County State Park to my office in Columbus. That would be my longest run to date. The *Columbus Republic* newspaper noted the event with a piece titled "Happy Birthday to the Long-Distance Runner." When I arrived at the office, the staff had erected a sign over the front entrance, "Finish Line. Happy Birthday, Boss."

So after one 10k race and the mini-marathon, I started thinking about the Boston Marathon. Can you believe it? I'd run two races in my life and now considered running that prestigious marathon. Why not? My running improvements had been amazing. First, however, to run Boston I had to qualify. The Boston Athletic Association required its marathon applicants to become eligible for entry by completing a marathon at a prescribed age-determined pace in the previous year. As a forty-year-old, my qualification time had to be under three hours and thirty minutes, an eight-minute-per-mile-pace. Boston was run in April, and, its already being fall, I needed to find and run another officially sanctioned marathon.

In December 1978, my wife and I drove to Huntsville, Alabama, for the Joe Steele Rocket City Marathon. I had trained for four months by running fifty to sixty miles a week and sensed that I was ready. I had prepared a wristband with each mile's split times carefully noted. Experienced marathoners told me, "Don't go out too fast, drink lots of water, and hang in there!"

Huntsville had beautiful weather on race day. Low gray clouds shrouded the sky, temperature was in the low fifties, and misty rain hung in the air — perfect running conditions! The race started outside the city in a park at the top of a hill. The first mile was all downhill. Without exertion or even realizing it, I picked up a minute ahead of my planned eight-minute-per-mile pace, but I considered it a gift and a good cushion should I need it. My running technique involved not looking ahead but focusing only on the immediate fifteen yards in front of me, almost like running in a hypnotic trance. At thirteen miles my wife met me with a small squeeze bottle of honey that someone had suggested as a secret elixir for energy renewal. At each mile marker, I checked my time and did a little skip for joy to see I was right on pace with that one-minute safeguard.

Then there was a moment of panic. Near the end of the competition, I somehow missed seeing the twenty-two-mile marker. My watch indicated it should have been where I'd just passed. Oh, no, I thought. I became distracted and lost the pace, and now I'm two minutes ... oh no, three minutes behind!

Fortunately, the twenty-three-mile marker appeared and confirmed I was still on pace. I was going to do it! Boston, here I come!

Finish at Huntsville, 1978

The Boston Marathon is run on the first Monday in April, Patriots Day, the celebration of Paul Revere's ride, which is a holiday in Massachusetts and one or two other states. My wife and I took our boys, Eric, Chris, and Brian, with us to Massachusetts. They would miss classes on Monday, but we planned to drive all night after the race to have them back in school on Tuesday.

The race starts in Hopkinton, Massachusetts, 26.2 miles west of the Prudential Center in Boston. The route passes through the towns of Framingham, Natick, Wellesley, Newton, and Brookline. It's an incredible event. "The Boston," as it is known to runners worldwide, is the oldest annual marathon, first run following the Greek Olympics in 1896. Why such an odd mileage, 26.2 miles? The answer is that in 1902 London hosted the Olympics, and the Queen wanted the race to finish under her balcony. That added two-tenths of a mile, and it became the standard for every marathon thereafter.

Twelve thousand runners lined up for the start in Hopkinton. My starting slot, determined by my qualifying time, was far back in the pack. I could see media helicopters hovering over the starting line a half-mile ahead but could barely hear the starting gun when it fired. There was a surge and then a stop as runners started to run but jammed into each other. It took fifteen minutes to cross the starting line, but even then, we jogged and stopped, jumped to the right, wiggled to the left, found space, sprinted ahead, and continued to dodge congested traffic. It took forty minutes before I could run a constant pace.

It was cool in Massachusetts that Monday. The initial miles of the racecourse became littered with windbreakers, knit hats, and gloves as runners shucked their wraps when body temperatures rose. Any observers wanting expensive Adidas or Nike running gear had their pick. Those contestants failing to wait in the long lines at the porta-potties prior to the race now simply paused beside the road to relieve themselves. They included a number of females, which was the first time I'd witnessed such openly immodest behavior.

At 4.2 miles in Ashland, the crowds increased and someone held up a sign that read "22 miles to Boston." Was that supposed to be encouraging? At Framingham, the crowds grew bigger yet. Everyone held out their hands to be slapped by the passing runners. I've never been told I looked so good by so many people in my life.

We passed through Natick at ten miles. That is an important point in the race, because there you assess your condition for the rest of the marathon. Being tired at that point signals trouble; being energized is good but not going too fast, you hope. I smiled seeing my elapsed time. I was on pace after subtracting the time it took to get to the starting line.

Passing through Wellesley and the campus of Wellesley College became the emotional high of the race. The women of Wellesley greeted us with an uninterrupted ovation — a scream, really — louder than any we would hear that day. How energizing it was; suddenly I was running more upright, bounding through that throng of coeds.

At sixteen miles, I passed Walt Stack, a seventy-two-year-old runner from San Francisco who runs with a can of beer in his hand. He's famous, thanks to *Sports Illustrated* and other sport magazines. The crowd knew him from the articles. I waved as I passed and then noted that this septuagenarian had outrun me for sixteen miles. Must be the beer.

All Boston runners are aware of the infamous Newton Hills — there are four of them. The first is a quarter-mile sweeping turn past Brae Burn Country Club; the second is another quarter-mile knoll, leading to the third hill, a sharp, wrenching eight hundred yards past Newton City Hall. The crowds during this stretch were very collegiate, and I found myself being offered beer continually. Walt Stack would say I should have taken one.

The fourth hill is the legendary Heartbreak Hill. It's not more difficult than the other three, but at 20.5 miles your reserves are about depleted. It's here that runners speak of hitting the wall. My legs were deadened, my energy waning, but I knew there were still six nasty miles ahead. As I neared the crest of Heartbreak, an Irish cop, well known for his presence at this spot, was there to spur us onward. "What a wonderful effort you have made, lad. You're looking splendid! It's only a bit further. Godspeed to you, son."

Wow, that put a few more coals on the fire!

When I crossed the summit of Heartbreak, I got my first sighting of the Prudential Tower. There were still several miles to go, but dammit, I knew I was going to finish.

The last stretch toward the finish line was a hazy mix of music, cheering, exhaustion, and pain. But I did it. After crossing the line, the staggering two-block walk to the family receiving area was trancelike, with strangers slapping my back, my muscles moving only by habit, but with the satisfying glow of accomplishment and pride. I just ran Boston!

I finished in three hours and forty-one minutes. Not bad, considering the slow start. The *Columbus Republic* had a feature story:

There were two follow-up articles about Bill Rogers winning the Marathon in two hours, twelve minutes, and Stark finishing at 3:41. Shoot, Bill had a head start.

Columbus' Jim Stark: He's a Marathon Man.

Chapter Five

Where's Brian?

My family needed to leave immediately after running the Boston Marathon to drive all night to get our three sons — Eric, Chris, and the youngest, Brian, age seven — back to school the next day. They had already missed one day of classes and needed to get home. An event occurred during that drive home that forty years later is still more memorable than my having run the marathon.

After the Boston event, I was emotionally pumped up. Although I felt physically tired, my mind buzzed with the color and excitement of the day. I could hardly wait to get home to share my experiences with my friends. I told my wife I would drive the first leg of the trip.

Our family had a new Dodge van, which had two bench seats in the back. In 1979 vans weren't nearly as fancy as they are today. The bench seats were mounted on steel legs with open spaces beneath them. Eric and Chris each had his own bench to stretch out on, while Brian, being the youngest, had to crawl under the second bench, where he slept near the rear floor heater, out of sight and mind.

When we reached the New Jersey highway, three hours from Boston, I pulled off the interstate at the Delaware Water Gap. Time to fill up the tank before the nighttime ride, and I announced to my passengers, "Okay, boys, last chance for the bathroom for a while."

Eric, Chris, and Brian obediently scampered off to the station's restrooms. As I filled the tank, I couldn't believe that even after running for nearly four hours, I was still feeling wide awake. I told my wife to get some shut-eye, I would continue to drive. After a visit to the restroom and buying a cup of coffee, we were off. I could hear the boys arguing for seat positions, but they soon quieted down, and I was left with my thoughts, reliving the race one more time as the miles disappeared behind us.

Six hours later, with the fuel gauge demanding another feeding, I pulled into a gas station in Ohio and again announced, "Okay, boys, everybody out. Go use the restroom."

And out went Eric ... and Christopher ... and ... where's Brian? Seven-year-old Brian wasn't in the car! What the heck ...! My God! We must have left him back in New Jersey! Mom became frantic! "In New Jersey? Oh no! What are we going to do? Didn't you see he wasn't in the car? Oh, my poor little boy!"

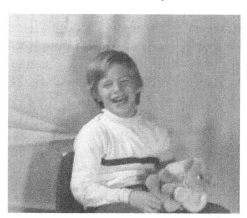

Brian, age 7

We spent several panicky moments on the driveway, looking for someone to blame: me, his brothers, the darkness, the big van. What to do? Who to call?

As shocked and horrified as we had become at such an alarming discovery, the good news was how quickly our minds were put at ease. Using the station's telephone, I called the Ohio State Police. "Hello. My name is Jim Stark, I'm in Blaine, Ohio, and it appears ..."

"Yeah, yeah. You left your kid in New Jersey, right? The New Jersey State Police have him, and he's fine."

The police gave us directions to the state police barracks at the Delaware Water Gap — a six-hour return — and I brought the tearful mother up to speed. Even in such a traumatic circumstance, my devious mind silently wondered what Mom would say if I suggested we have Brian shipped home via UPS, but I kept that sick joke to myself.

Brian was not at the station; he had gone home with one of the policemen whose shift had ended. The trooper lived on a farm only a short distance away. When we arrived, Brian was playing with some ducks. He didn't react with excitement upon seeing us, despite being smothered in hugs and kisses by his mother. "Brian, are you all right? We're so sorry. What happened? Brian, we were so worried."

"I'm fine."

"But what happened?"

"You left me. I tried to catch you, and this man and lady saw me, and the police came, they were nice, and I came here. He has goats, and sheep, and ducks."

The story, once we pieced it all together, was that Brian walked out of the gas station as we were leaving. He called out to us and ran a short distance until we disappeared. A couple pulling in at that moment saw the whole incident unfold and went to Brian immediately. The couple didn't give chase because they couldn't tell which direction we turned when approaching the highway. Almost immediately, a New Jersey highway patrolman pulled into the gas station. The couple explained that Brian's parents drove off and left him.

"They'll probably return shortly, when they realize he is missing," the policeman said, and he began asking Brian questions. We had prepared our son for just such a circumstance, and he was well rehearsed.

"What's your name, son?"

"Brian R. Stark."

"And where do you live?"

"4320 North Washington Street, Columbus!"

"Is that Columbus, Ohio?"

"No, just Columbus." To which they surmised that it must be Ohio.

Apparently, we had not emphasized to Brian that he lived in Indiana. I later recalled that after leaving the gas station in New Jersey, police cars had slowed down when coming up behind us, then zoomed by as they looked for a white van with Ohio license plates.

Brian was back in class the day after we returned, and his story soon became known by the entire school. That night, with all of us at the dinner table, my wife asked, "Brian, now that you're home, and you've talked about your experience, how do you feel about it?"

Brian thought a minute, then said, "I'm proud."

And I thought that's good. For a seven-year-old to experience such a potentially terrifying situation, but to discover that strangers can be kind, policemen are friendly, and even the most frightening event can have a happy ending, is a good thing.

The story of us leaving Brian soon became the talk of the community, and we heard from everyone, it seemed. The calls didn't castigate us for being bad parents, but rather told similar stories about almost doing the same thing.

My favorite story involved a family taking their ninety-year-old mother south for a spring vacation. As they were leaving town, they stopped to fill up the tank. Mother used the restroom, and the family drove off without her. Five minutes down the road, they realized their oversight and sped back to the gas station just as she walked out of bathroom.

"Okay, Mom, time to get going," they said. Not another word was spoken.

Another family reported that, at dinner one night, a mother told her children Brian's story as a lesson of what to do in a similar circumstance. Suddenly two of the younger kids began crying mournfully, "Don't leave us. Oh, please don't leave us!"

Years later, Brian, in addition to becoming a schoolteacher, had become a long-distance runner, as in running across the United States in 1998! In 2011 he was asked by the director of the Nevada State Parks to set a record by running five hundred miles across Nevada on trails, and climbing fifteen mountain ranges, and to do it in ten days, fifty miles a day! Near the finish line at Lake Tahoe, he was interviewed by a Reno television station. His mother and I had flown out to Nevada to see him finish and were standing behind the television interviewer. "Just what makes a person want to run such incredible distances?" he asked.

"Well, I don't know. My parents drove off and left me in New Jersey when I was little, and I've been running ever since."

Brian, 2011, TV interview

# # #

Chapter Six

Setting Sail

I was standing watch the day we sailed into large groundswells, non-breaking waves that reached ten to fifteen feet in height from low point to crest. The *Brilliant*, a sixty-two-foot, two-masted schooner, lifted twenty feet or so in the swells, then plunged into the water's surface as it fell into the trough. I stood at the end of a bowsprit, a fifteen-foot spar extending forward of the bow. The bowsprit served to secure the headsails, but there was a circular railing at its end where the bow watchman could stand, keeping a lookout for other boats and floating debris.

At first only my feet got wet, but as the swells grew in size, I started holding my breath as water rose to my knees on the downward plunge. I was having great fun until the captain saw me. "What's that kid doing out there!" he shouted. "Get him back on deck." It was 1953 and I was fifteen years old at the time.

Scouting got me into sailing. I began as a Cub Scout, then transitioned into Boy Scouts, and Explorer Scouts became the next step in the program. The Explorer program was available to boys fourteen through eighteen. There were units called Air Explorers and Sea Explorers, specializing in flying and sailing respectively. An experienced yachtsman in our community named Bill Workman suggested the idea of forming a Sea Explorer Post and volunteered to be its leader.

Sea Explorer uniforms were patterned after regular Navy enlisted uniforms. We had the thirteen-button, bell-bottomed trousers

as well as the standard blue Navy jumper, referred to as "crackerjacks" after the image appearing on the Crackerjack box. The only difference was that the regular Navy uniform had three white stripes on collar and cuffs, whereas the Scout uniform had two stripes. And just like Navy sailors, we wore the white "Dixie cup" caps. We thought we were pretty cool.

Bill knew of a sailboat in need of repairs. We acquired it for free, and our initial meetings involved working on the boat, getting it ready for the water. Although two of our fellow scouts had sailed on Connecticut's Long Island Sound before, most of us had never been on a sailboat.

That summer, while our boat was being refitted, Bill learned that Mystic Seaport, a historic maritime organization located in New London, Connecticut, offered a weeklong offshore sailing school for youth organizations. The youngsters served as the ship's crew. Bill made the arrangements, and eight of us enthusiastically signed on for what promised to be an exciting experience.

The boat we would be sailing, the *Brilliant,* was famous. Designed by the renowned yacht design team of Sparkman and Stephens, it was built in 1932 as an ocean-racing yacht. On its maiden voyage, the *Brilliant* set a world record by crossing the Atlantic Ocean in only fifteen days.

Brilliant, 1953

The Coast Guard acquired the schooner during World War II and patrolled the New England coast on the lookout for submarines. During that period it had two machine guns mounted on its deck. After the war the *Brilliant* was purchased by Briggs Cunningham, known for both auto and yacht racing. Cunningham donated the boat to the Mystic Seaport in the early 1950s, around the time our Sea Explorer unit went aboard.

In addition to our leader, Bill Workman, we had a captain, Mike Jones, and a first mate, Ralph "Rusty" Nale. Serving as the crew, we eight Scouts stood watches while under way, which involved spending two hours at the helm, two hours standing bow watch, and two hours working in the galley.

We sailed out of New London, Connecticut, across Long Island Sound to different ports on Long Island. Although we ventured out into the Atlantic, we never traveled far from shore and never sailed overnight.

    Captain Mike was great with young people, and a wonderful instructor, teaching us nautical terms and sailing techniques and strategies. The captain held master's papers to command anything under sail regardless of its size. Rusty spent most of his time in the galley overseeing the preparation of our meals and showing us, his galley slaves, what needed to be done. Rusty had a delightful sense of humor and invited as much kidding from us as he dished out. "Okay, my laddies, those pots better be clean and bright, or I'll be taking you for a ten-foot walk on an eight-foot plank."

    With all the chores completed in the evening, Rusty would entertain us on deck by telling stories of rich yachtsmen he crewed for while sailing around the world. I thought that would be the most exciting adventure I could imagine. I enjoyed Rusty and I could tell he took a liking to me as well. One afternoon with just the two of us washing dishes, Rusty said, "You like it out here on the water, don't ya, Jimmy?"

    "Yeah, I love it."

    "Thought so. I know some boat owners who sail out of these waters each summer, looking for crew. They sail from here to Bermuda, then down to the Virgins. Usually back in the fall before the hurricane season. Would you be interested in doing something like that, Jimmy?"

    "Holy cow. That sounds great!" Of course I couldn't imagine what my parents would say, but Rusty gave me his contact information and told me to get in touch with him after our cruise. I told one of the other boys, Bruce, about Rusty's invitation, then Bruce blabbed it to everybody else. I was the only one Rusty had talked to regarding that opportunity.

    We went ashore several times during the week. We boys would saunter into town like crusty old tars, acting like we'd just returned from sailing around the Horn. One of the guys, Peter, bought a paperback romance novel with a sexy lady on the cover showing her long legs and barely covered boobs. He found some racy passages in the book describing her sexual romps, and in the evening we'd gather on the deck as Peter read aloud, with all of us hooting and hollering at the vivid images those readings conjured up.

What a memorable summer experience it had been, one that presaged other sailing adventures throughout my life. My mother, however, always considered that summer a bad influence. She picked several of us up at the Bridgeport train station when we returned from New London. She was shocked at the foul language I had become accustomed to using. "Hell" and "damn" seemed to punctuate every sentence as I described my wonderful week to her.

"Jimmy, what has become of you?"

"Home from the sea, ma mère, home from the sea."

I mentioned Rusty's invitation to my mother some weeks later, and it became clear she and my dad would not approve. I had just started prep school the previous year and was working with a tutor that summer to try to improve my grades in Latin and French.

***

In 1963, a year after graduating from college, I found myself in Norfolk, Virginia, as a Navy lieutenant—junior grade, having recently received my wings as a Navy pilot. Before I checked into my new squadron in Brunswick, Maine, I started six months of specialized training, the first of which was Nuclear Weapons School in Norfolk. The school was top secret and allowed no materials to be taken off base. That meant no homework, so our evenings became free to spend as we wished. For most of my friends, it meant nights at the Officer's Club, but I spent the evenings at the base hobby shop building a sailboat.

The recreation department at NAS Norfolk had a well-equipped woodworking shop. It had all the hand tools a woodworker could ever want, plus power lathes, drill presses, bandsaws, and circular saws. All were available for the use of base personnel. I would be in Norfolk for six weeks and assumed that would be enough time to build a simple sailboat. A sunfish is not much more than a surfboard with a sail. Fourteen feet in length, it has a shallow cockpit, a daggerboard, and a single sail. The sunfish was constructed with an internal framework of marine plywood between its deck and hull. I immediately went to work. My bride, married less than a year, spent her evenings perched on a sawhorse, watching me work. Not much fun for her, but she was a good sport about it. I completed the project just as my assignment ended in Norfolk. My next orders had us heading to navigation school in Jacksonville, Florida.

We were a two-car family; my wife drove a Chevrolet Corvair convertible she received as a college graduation gift, and I owned a '57 Volkswagen Beetle I had before we were married. The thirteen-foot sunfish mounted on the roof of the VW, hanging over the front and rear ends of the car. I was quite a sight traveling down the highway.

The sailboat's maiden voyage took place on the St. Johns River in Jacksonville. The St. Johns is 310 miles long, with its origin in central Florida. It flows north and empties into the Atlantic at Jacksonville. In some places the river is three miles wide and is used by ocean freighters and tankers. Naval Air Station Jacksonville had a beach on the river where we launched the sunfish.

Our little boat had a displacement of four hundred pounds and barely supported two persons; bathing suits were required. From shore, you saw two people apparently sitting on the water's surface, propelled by a sail.

On the maiden voyage, after thirty minutes of excellent sailing, I discovered my wife and I had gone more than a mile out into the river. Looking back to where we'd parked the car, I was shocked to see the shore obscured by dark clouds and sheets of rain. We wouldn't be able to return in that direction.

Looking downriver, I saw that the distant shore looked somewhat clearer, and I could make out large homes along the water's edge. When the squall hit, it pushed us toward those homes. We were flying. Sharon, lying flat on the deck and holding onto the mast, was screaming. I slipped over the side to act as a drag to try to keep the boat upright. With one hand holding the deck's handrail and the other the sail's mainsheet and tiller, my biggest problem was keeping my bathing suit on! We were racing through the water, and I drew up my knees to keep my trunks from being pulled off. As we approached shore, we could see two people at the end of a pier watching our struggle.

They were a middle-aged woman and a young man, about my age. The woman, I learned, was an admiral's wife, and the young man the admiral's aide.

Congratulations, Lieutenant Stark, you just beached your vessel on its maiden voyage smack in the middle of Admirals Row!

The admiral wasn't home, but Mrs. Admiral was most gracious. She draped my shivering spouse with a bath towel, ushered us into the house, and ordered hot tea. My wife was a good talker. "Oh, you are so wonderful. My husband is a Navy pilot, and we are new to Jacksonville. We just love it here, and we're so sorry to intrude. You are so kind."

The storm passed as quickly as it arrived. I figured it was just a short sail upriver to our car. Sharon could hold her own with Mrs. Admiral while I sailed back to the car, loaded the sailboat, and returned for my shipwreck survivor.

The sunfish had one sail and was not good about tacking into the wind. The short trip back to the car took two hours. I said my wife was a good talker, but this really put her to the test. The experience apparently exhausted her repartee, because there was nothing but silence in our house for days afterward.

We did have some pleasant sailing experiences on the St. Johns following that initial cruise, so our first voyage was not a game-ender.

We spent six months in Jacksonville, first at the navigation school, and then, still in Jacksonville, I was ordered to Training Squadron Thirty, where I received flight instruction in the P2V Neptune, the aircraft I would be flying in the squadron.

That phase of training went well. At its completion, it was off to Brunswick, Maine, where I would join Patrol Squadron Twenty-one, for my three-and-half-year tour of duty. It was time to load the sailboat and drive to Maine.

On other transits with two cars, before the era of cell phones, we used hand signals for communication on the road that I learned in formation flying. A thumb in your mouth meant the need for gasoline. A finger circled over your head meant to rendezvous — let's stop and discuss plans. My bride wasn't enthusiastic about these methods but grudgingly played along.

Leaving our rented home in Jacksonville, I, with sailboat strapped to the top of the VW, and bride in her Corvair with her new puppy, Happy, as a passenger, we dropped the keys through the locked front-door mail slot and started out. There were two ways to go from our house to the highway, although one was shorter and the obvious first choice.

I left second and didn't see my wife but assumed she was just ahead of me. When I got to the highway and still didn't see her, I decided she had gone the long way, so I waited. And waited. Did she have a problem? Was the dog sick? Where the hell was she? She certainly wouldn't have started down the highway without me behind her. I thought I had better go back home and check. When I did, there was no sign back at the house, so I returned to the highway. And waited. What to do? It had now been over an hour since I left the house. Perhaps she did go ahead on the highway. She had to be as concerned as I was. Who would she call? Her father in Lafayette, Indiana, might be her first choice. So I called him. "Hello, sir, this is Jim in Jacksonville. I lost your daughter."

"Huh?"

After explaining the situation, we decided I should drive for three hours to Savannah, Georgia. There, find a Howard

Johnson Restaurant — every town of any size had one — and call him to see if he had heard anything.

Driving out of Jacksonville, I passed an Esso station. Sitting between the pumps watching the passing traffic were my bride and Happy. Seeing them, I swerved into the station, and observers would have thought we hadn't see each other since World War II.

She had indeed gone the long way from our home out of habit and then, as I did, waited and went back and forth searching for me. Finally, she pulled into this filling station and tearfully approached the attendant: "I've lost my husband. He's a Navy pilot and he's gone."

"When did he die?" he asked.

"No, not dead. Lost. An hour ago. Driving a Volkswagen with a sailboat on top."

"Err, sailboat on top, huh? Well, little lady, why don't you wait right here. I'll call the state police. Shouldn't be hard to spot a car with a sailboat on top."

My 1957 Sunfish Transportation System

# # #

Chapter Seven

ABATE

Early in my motorcycle riding experience, I took a motorcycle survival course in Lafayette, Indiana, taught by the ABATE organization. ABATE is a group of mostly Harley-Davidson riders who do good things in offering safety seminars and skill training for the motorcyclist. Their instruction is excellent. Participants who had ridden for years exclaimed that they learned techniques and skills they never knew before.

ABATE members are vehemently opposed to legislation such as mandatory helmet laws and have been very successful in their opposition. ABATE and the National Rifle Association have a lot in common in their political clout. ABATE publicly stands for "A Brotherhood Aimed Toward Education." However, it secretly means "A Brotherhood Against Totalitarian Enactments." I got to know the ABATE instructors. Good guys all, although at first sight, the leather, menacing-looking tattoos, and pierced body parts tend to scare off timid souls. And for the record, I do not support their opposition to mandatory helmet laws.

A few weeks after my riding course, the Bean Blossom Boogie hosted a week-long motorcycle gathering in Bean Blossom, Indiana. I heard it was a rather wild affair and, curious, took my thirteen-year-old son, Brian, on the back of my Honda and rode over to Bean Blossom. Not wanting to pay the fifty-dollar weekend fee, I hid the Honda in the bushes beside the road and walked in asking for Zach, an ABATE bigwig whom I met at the survival course.

I told Brian, "There're going to be some shocking sights in here. Just look straight ahead. Be cool. And for gosh sake, don't tell your mother."

Brian wore wrap-around sunglasses, which I thought looked pretty dippy, but they hid his eyes, which I'm sure worked overtime taking in the sights.

At the entrance to the gathering, motorcyclists were rolling in, having their wrist bands checked by a volunteer. Most of the bikes were Hogs, the nickname for Harley-Davidsons. Almost all of the leather-clad, helmetless riders had a babe on the back seat with a case of beer on her lap. Someone had found a discarded couch by the roadside and dragged it near the entrance. On it sat four dudes, their feet buried to the ankles in empty beer cans. Dust from the arriving traffic covered their heads, arms, and legs. One of the men had a sign. As each biker and passenger passed, he would hold it up: "SHOW US YOUR TITS." And most of the women did.

"Just look straight ahead, Brian!"

I found Zach, and we had a brief chat. "Isn't this something?" he said.

Yes, indeed, it was something.

That 500cc Honda was the first of several motorcycles I owned. A new 900cc Yamaha became my next bike, and then over the years, I owned three Honda Goldwings. The first Goldwing was a rebuilt, used, 1984, 1200cc bike. The motorcycle had been damaged in an accident but was being rebuilt by a man in Brazil, Indiana, who owned a Honda shop. I saw the half-finished machine and knew it was going to be as good as new when completed. I wanted to buy it but didn't want to commit until making a final inspection.

Remember how unsupportive my wife was about motorcycles? Needless to say, I didn't mention my interest in buying the Goldwing to her.

When I heard he had finished the motorcycle, I asked my friend John to fly me in his airplane to the Brazil airport, where the seller would meet me and drive me to the garage to inspect the rebuilt bike. I explained to John on the flight over that I was sure I would buy the machine but didn't want the seller to see him fly off and know I was stranded in Brazil with no choice but to make the purchase. So, with a wink, I told John, "When we arrive and within earshot of the seller, I'm going to say to you, 'Wait here until I let you know if I bought the motorcycle.' But since I'm sure I'm going to buy it, after we drive off, you can leave."

The Goldwing was beautiful, absolutely brand-new looking. I wrote a check and rode off to my office warehouse where the bike would be kept. When I got home and walked in the house, my wife said, "Did you just buy another motorcycle?"

"What? Geez! What makes you think that?"

"Because John just called, and he's been waiting for you for two hours at the Brazil airport."

Oh, John, why didn't you understand!

I bought that motorcycle for $5,000, kept it for five years, and sold it for $4,500.

In between the 1984 Goldwing and my next purchase, I was divorced. No, it didn't have anything to do with motorcycles. She was deputy-mayor of our city and campaigning for the mayor's job. We just drifted off in different directions. She's a very talented woman and is an incredibly good mother for our sons. We have an excellent relationship.

The next Goldwing was a new 1992, six-cylinder, 1500cc model I bought for $9,000, kept for twelve years, and sold for $6,500.

When looking for the 1992 machine, I called dealers all over the United States, asking for their best deal. I didn't care where I found the lowest price; I'd travel there and ride the motorcycle home. Charlotte, North Carolina, had the best offer. My new sweetie — this was several years after my divorce — an enthusiastic motorcycle passenger, and I, drove down, got the bike, then spent a couple of days in the Smoky Mountains before heading home.

I purchased the third new Goldwing in 2004. I thought I would keep the 1992 machine forever, but I made the mistake of test riding a 2004 model and just had to have one. It was a new design with a more powerful engine, and so perfectly balanced you would think it had training wheels.

I went through the same best-deal search from Maine to California and found the lowest price in Akron, Ohio. No problem, it was only six hours away by car. I was living in Bloomington at the time, so my new wife and I planned to drive over to pick it up. On the way I told her about the Goldwing's new features, one of them being an air vent in the middle of the windshield.

When we arrived, the dealer said he had my purchase serviced, installed the luggage rack I wanted, and the bike was in the middle of the showroom floor. And there it was, a gleaming black beauty. My first Goldwing had been dark brown, the second, deep maroon, and this one, the blackest black, to show off its chrome accessories. Other customers in the showroom also admired the bike.

Then my wife said, "Didn't you say it had an air vent in the windshield?"

"Yes, I did. What's the deal here?" When I pointed this out to the salesman, he stammered with a stunned look on his face, then ran off to check the paperwork and to confer with his boss. Oops, this was not the 2004 I wanted, but a 2003! They mistakenly got this one ready in error, even adding the luggage rack. I'm giving them the benefit of the doubt about it being a slip-up.

They had no 2004s in stock. They talked about lowering the price on the 2003, but I wanted a 2004. They offered to find one and figure out some way to deliver it, but I said no and accepted their coupon for a free lunch. We drove six hours back to Bloomington.

Evansville, Indiana, had the second-lowest offer, so a week later we made another five-hour drive and took possession of a beautiful new 2004 Honda Goldwing.

2004 Honda Goldwing

My motorcycle has a name. I held a contest among my VIPs to suggest one. The winner was Paul Arnold, who suggested "Rocinante," the name of Don Quixote's horse. Rocinante was a stallion, and it's appropriate that my hard-charging six-cylinder, two-liter motorcycle is male. I call it Rosey, like the famous NFL football player, Rosey Grier, the 284-pound, six-foot-five, original member of the Fearsome Foursome of the Los Angeles Rams.

#

Chapter Eight

Motorcycle Buddy

It was 1989. The day dawned gray and cold with a promise of rain as Brian and I set out from our home in Columbus, Indiana, on our third spring break trip together. Brian rode shotgun in the back seat of my Honda Goldwing motorcycle. He had become my motorcycle buddy.

Many of my adventures have included my three sons, either individually or together. Lest you think I'm running for a father-of-the-year award, let me confess the selfish nature of our time together. Yes, selfish because, first, our activities involved doing the things I loved to do: camping, flying airplanes, riding motorcycles, and running. And, second, I was in charge, I made the rules, and I was the boss! You don't have that kind of authority when on an outing with friends.

One of the control gimmicks I used with my sons was a whistle I hung from the rearview mirror of the car. I explained to the boys I expected a certain amount of roughhousing in the back seat during our trips. I could tolerate some degree of horseplay, shoving, and arguing; however, when it became too much, I'd blow the whistle and then expected quieting down and obedience.

Invariably, at some point during our outings, the boys would begin to get loud, which led to boisterousness, but before I could take action, the noise would stop, and I'd hear, "Are you about to blow the whistle, Dad?"

"No, not yet, but you're getting close."

Amazingly, I never once had to blow that whistle.

Another gimmick that seemed to work was report cards. I let the boys know before our trips that they would be receiving report cards with grades for helpfulness, behavior, punctuality, neatness, and cooperation. Some years later, after my middle son, Chris, had entered high school, I noticed he still had those report cards displayed on his bedroom wall. When I asked about them, he said, "Heck yes, those are the best grades I ever received."

One of our annual trips was a week in Wisconsin, the first few days spent at the annual Experimental Aircraft Association airshow in Oshkosh, and the remaining days camping in a Wisconsin state park. Some of those outings involved flying to Oshkosh in our airplane and camping under the wing, while in other years, we towed our pop-up camper.

The annual trips first started when Eric, the oldest, was just eight, and Chris six. Brian, a two-year-old, was considered too young to go. His exclusion from those early outings was understandable, I thought. The year Brian became five, while I was planning the week with Eric and Chris, Brian asked, with tears in his eyes, "Can't I go? I don't suck my thumb anymore."

"You bet, Brian. This is your year!"

The boys and I had uniforms. I had designed a logo, composed of a star and the letter K, which when pronounced said "STARK." I used to build radio-controlled model airplanes and decorated the models with that symbol. I had the logo made into patches sewn onto ball caps and shirts. I don't know what people thought seeing our foursome walking the grounds at the airshow — a singing group maybe — but it did help me find the boys in the crowd.

L to R: Chris, Eric, Brian (kneeling), 1975

Nineteen eighty-six was the year I flew across the United States with Eric, and also the year I ran across Florida with Chris. It wasn't a conscious plan to do both the same year; they just happened coincidentally. Brian was thirteen at the time.

Brian and I used to fish together when he was growing up, and it always astounded me how patient he was, sitting in the boat waiting for fish to bite. After a long day with nothing caught, I might say, "What do you think, Brian? Time to call it quits?"

And he would reply, "Just five more minutes, Dad." Amazing.

One summer afternoon in 1986, while in my office gazing out the window, I decided it was a good day to play hooky. I called Brian, home after school at that hour, and told him to round up the fishing gear. "Let's go catch some fish."

A friend had a farm pond full of large bass and crappie. We spent a delightful afternoon hooking into some beauties. Rowing back to the pier, Brian asked, "Is that it, Dad?"

"Errr, what do you mean?"

"Is that it? You and Eric fly across the country, and you spend a week running across Florida with Chris. And me you take fishing. Is that it?"

"No. Not really. Errr. I've been working on a plan but wanted it to be a surprise. I'm not quite ready to tell you about it yet but will soon."

He bought it — at least I think he did. A couple days later I told Brian that he and I would fly down to Gatlinburg for several days of fun at Dollywood, Pigeon Forge, and Smoky Mountains National Park.

Never let it be said your children don't keep score on who's getting special attention. Actually, Brian may have gotten more than his share, because he became my motorcycle companion on several spring trips. We traveled to North Carolina the first year, then down to Pensacola, Florida, the following year. In 1989 we set our sights on New Orleans.

Interstates were the less preferred roads on our trips, but with cold weather and rain threatening, we wanted to get out of

Indiana's early-spring coolness and into the moderate temperatures of the South as soon as possible. When we entered Kentucky, we left the super slab and found the winding ways of rural back roads.

From Brian's journal:

After crossing the Ohio River, we entered the world of real people. It was at a restaurant in which we stopped to thaw-out. The real people were there to eat, socialize, and have a good time. We were there to drink hot chocolate and regain the feeling in our fingers and toes. We discovered an electric hand dryer in the men's room that when aimed at the unbelted open top of our trousers felt wonderful.

My new electric vest and gloves were a godsend. Also found later that day, with seemingly heavenly guidance, was a motel in Nashville, Tennessee, with an in-room Jacuzzi. Wow, did that hit the spot.

Nashville's Grand Ole Opry was something I'd wanted to see for some time. Rather than fight the Saturday-night traffic, we opted to sign up for a tour that picked us up at the motel and brought us back after the show.

One of the interesting stops the tour bus made en route was at a truck terminal. It's apparently common for truck drivers coming into Nashville to park their rigs at the terminal and take in the Opry as part of their evening's entertainment. Fifteen truckers of varying sizes, shapes, and genders scrambled aboard. Apparently, mother truckers are no longer a rarity behind the controls of today's eighteen-wheelers. The conversation of that merry band proved them to be a proud lot. All comments related to truck driving in some connection. In the space of twenty minutes, Brian and I learned: J. B. Hunt drivers were impatient; Greyhound buses were "puppy dogs"; a forty-eight-foot flatbed truck can get into any side street; and pets in the cab include not only dogs and cats, but even bobcats!

Brian:

The truckers shared their love for large vehicles with everyone within hearing distance, and cautioned our tour bus driver about his driving. After all, he had fifteen professionals in the back watching very closely.

We had a rip-snortin' good time at the Opry! The next day, a father-and-son jog through downtown Nashville gave us a real feel for Music City. We toured the hill on which Nashville sat, ran by the Ryman Auditorium, where the Opry first started, and saw recording studios, instrument stores, and numerous other music-related businesses.

Our Goldwing bade adieu to Nashville and carried us southwest toward Columbia, Tennessee, where we rode onto the Natchez Trace. The Trace is a historic national park highway that runs 450 miles from Nashville to Natchez, Mississippi.

Brian:

The Trace has a 10,000-year history of use by Indians and early settlers. Pioneers used the trail as their route home after floating their goods downstream on the Mississippi River to trade at Natchez. Well-kept points of interest along the roadside included Indian burial mounds, waterfalls, sections of the original trail, and beautiful overlooks.

A mid-afternoon break near Waynesboro, Tennessee, found us dismounting in front of a small diner. The parking lot was empty, but the sign on the door said "Open." Taking the usual few minutes to unbuckle helmets and unplug the sound system, we were about to enter the restaurant when I remembered I wanted to look over a map as we ate. As I turned to retrieve the forgotten item, the proprietor burst from the front door and shouted, "You're not leaving, are you?"

It was a great lunch: fried chicken, black-eyed peas, greens, homemade biscuits — a real southern meal. Not bad for a California restaurateur who just moved to Tennessee. He and his wife had not been on the Trace and wanted to know all about it.

The Trace is a two-lane parkway, built in the 1930s, that twists and rolls through beautiful scenery. Its historical points of interest are fascinating. If it's your thing to ride a motorcycle, this is the great road on which to do it. We ended the day about mid-Trace, near Tupelo, Mississippi, an industrial center where many motels and restaurants cater to modern-day travelers and business people.

Brian:
The Tupelo motel had a sauna. Wanting to cleanse my body of its impurities, I jumped in and was sweating up a storm. Dad couldn't stand the heat, but just before he headed back to the room, reached in with a large cup of water and threw it on the hot lava rocks, saying, "See Ya." The steam was really hot. Thanks a lot!

The next day's ride to Natchez was challenged by on-and-off rain. One interesting nature walk along the parkway was the Swamp Walk at mile 122. It's a boardwalk that meanders through the cypress and lily pads. The stroll through the dark green foliage took twenty minutes and was a lovely, unique experience.

We left the Trace and entered the city of Natchez — along with several thousand other tourists. Spring in Natchez is the time of the annual Natchez pilgrimage. Confederate balls, parades, tours of historical homes, and nightly performances are just some of the activities taking place in this city in March. Brian and I got the last motel room in town — literally.

The next morning, we awoke to another rainy day. Route 61 to New Orleans passes through several beautiful southern towns and lush countryside. The discomfort of riding in rain suits and the struggle to keep bifocals clear of rain drops notwithstanding, the magnolia blossoms beside the highway were a sight to behold in their brilliant colors.

As we approached New Orleans, the precipitation increased in intensity. Just seeing the road became a challenge. With visibility down to nil, we called it quits twenty miles from downtown and dove — figuratively — into a Holiday Inn.

After drying off and warming up, we made plans for the evening. I wanted Brian to have the cultural experience of spending an evening on Bourbon Street in the French Quarter, so it was off to the Big Easy.

Bourbon Street was much as I remembered it from other visits over the years. Authentic New Orleans jazz still oozed from the doorways of French Quarter bars and clubs, people still strolled (staggered) arm-in-arm down the center of the street, drinks-to-go stands flourished to keep spirits high (pun intended), and strip joints and B-girls abounded.

Brian:

If you haven't met your dream-girl yet, she's waiting for you in New Orleans. Just remember to bring your MasterCard.

The next morning a quick peek through our room's curtains revealed a depressing sight. Heavy rain pounded down from leaden skies and flooded the motel parking lot. The thought of donning our rain gear, still damp from the previous day, and playing bumper tag with the early morning traffic, wasn't heartening. And the forecast was for twenty-four more hours of rain. We didn't have the time to lay over, so we needed to grit our teeth and move on. Perhaps, we hoped, our northwest route to Vicksburg would take us out of the dreariness.

It didn't. A long, wet ride of 250 miles ranked right up there with a trip to the dentist as a fun thing to do on your vacation. On the brighter side, our arrival in Vicksburg did coincide with a momentary break in the rain. We could at least reconnoiter the historic antebellum community, select our accommodations for the evening, and make arrangements for a tour the next day.

The Vicksburg tour was fascinating. The Civil War battle of Vicksburg came alive for us with vivid reality as we passed the trenches, batteries, and battle lines where 150,000 Americans struggled for forty-seven days to kill one another. Modern-day warfare is almost more palatable, with its long-distance killing, than standing fifty feet apart, firing musket balls into the flesh and bone of your fellow countrymen. Few soldiers survived even a minor wound in the battle. The meager diet, unfathomable filth, and rampant yellow fever took a heavy toll.

Vicksburg also had the remains of one of the Civil War's ironclad warships. The USS *Cairo* was sunk during the war in 1862 and lay preserved and hidden in Mississippi's silt until 1963, when it was discovered and raised. Everything from cannons to toothbrushes was recovered in the same condition as when sent to the bottom.

We got away from Vicksburg at 3:00 p.m., determined to reach Memphis before day's end. The weather was cloudy and cold, but not raining. When I put on my long underwear that first day in Columbus, I had no idea I would be spending the entire week in those long johns!

It was a hard ride, but by 8:00 p.m. we pulled into Elvis' hometown, ready for the first motel in sight. Our bow-legged waddle gave mute testimony that we were ready to be off the Goldwing and into a bed.

We decided before we left Memphis the next morning to "do" Graceland. Signs, brochures, and billboards insisted we do so. Graceland is not hard to find. Just take Elvis Presley Boulevard a few blocks south of the inner-belt, then stop at the home with the wrought iron guitar players on the gate and two jet airliners parked in the driveway.

The attraction is very commercial and might even be considered tacky. But regardless of what you think about the king of rock 'n' roll, he was indeed a phenomenon in the entertainment world. He had over 228 gold records, 48 of them since his death! Over one billion total records sold. He is truly a legend. We're glad we visited this shrine.

Leaving Memphis at midday, we enjoyed our first look at partly sunny skies in over a week. Our route was northeast to Kentucky Dam Resort Village, just a few miles from Paducah.

Brian:

The sun was shining, the tape player was blaring, and our noses were sun burning. It was a terrific day!

Having called ahead to make Kentucky Dam reservations, we were pleased to find that we were given a three-room cottage. The added space was appreciated. Getting close to your son was wonderful, but riding piggyback for seven days was close enough. We both enjoyed spreading out, if only for a few hours.

The sun came out on this last day of this trip, bright and shining. Clear skies with temperatures in the upper sixties were forecast.

After an early-morning run around the resort grounds, we enjoyed a large breakfast buffet at the lodge. One last souvenir stop in the gift shop assured Mom we hadn't forgotten her. Whew, almost! Then onto the wing for the ride home.

In spite of the rain, it had been a fun week. Brian and I had lots of jokes to snicker about when we relived the events of our trip. We'd even use the weather to prove our ruggedness as we embellished the challenging nature of our adventure.

Brian:

The spring break may have been the year of rain, but it was also the year of tourism. Spending time in each of the cities we visited was definitely an enjoyable experience. So many different places and sights to remember. And the Goldwing was great, even in the wet weather.

What a wonderful machine the Goldwing is. Isn't it interesting that after covering two thousand miles in eight days, I still took the long way through the neighborhood when returning home?

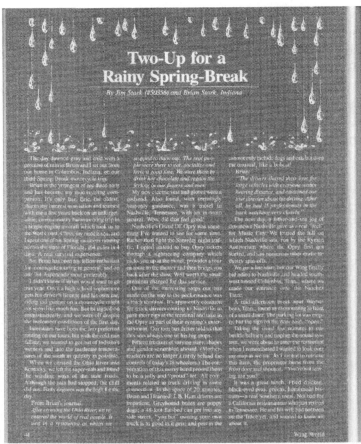

The story about our New Orleans trip appeared in a 1989 issue of *Wingworld* magazine

# # #

Chapter Nine

Flight Training Excitement

In the Navy we used to describe flying as hours and hours of boredom mixed with seconds of sheer terror. That's a humorous exaggeration, because "boredom" is not a word I associated with flying, although fourteen-hour patrols over the North Atlantic could get wearisome. But terror? Well, we stoically referred to them as "hairy moments."

Aircraft carrier landings set Navy, Marine, and Coast Guard pilots apart from the other branches of the military. Although Navy pilots are rather cavalier about that capability, believe me, every carrier landing is taken seriously.

Thomas D. Callahan III, or TD as we called him, had an interesting experience when making his first carrier landing. TD, two others, and I were scheduled to fly our carrier landing qualifications together. We would fly out in a four-plane formation.

We practiced carrier landings for two weeks beforehand at an outlying airfield with a runway that simulated a carrier deck. The field had the mirror landing system like the one on board the ship we used to visually keep us on the glide slope. What it did not have were arresting cables. On the carrier five cables stretched across the deck that, when snatched by the plane's arresting hook, brought the aircraft to a stop in fewer than one hundred feet. The first time we would experience that sudden halt was when we made our first actual arrested carrier landing. On the day of the qualification, we flew out solo. The back seat had no instructor to ensure it was done safely. Some students had never seen an aircraft carrier before the day they landed on one. This was just another of my often-asked questions during flight training: was I really ready for this? So far, like surviving the first night solo, formation flying, instrument flying, and aerobatics, I had always made it safely and confidently. The Navy apparently knew more about my proficiency than I did. Carrier landings were the last stage of basic flight instruction before we moved on to advanced training.

Callahan was an excellent pilot who had his private pilot's license before entering the Navy. Besides being a skilled aviator, TD just had a cool appearance, a casual aloofness that radiated professionalism, confidence, and a "hot shit" demeanor.

We were excited that day when taking off to fly out to the carrier waiting for us in the Gulf of Mexico. Nervous stomachs ruled the day. We tried to hide the sweaty palms and pale complexions behind a feigned bravado. No need for such acting for the cool Mr. Callahan, however. The three of us, flying in formation on Callahan's wing, practically puking in our laps, were amazed to see ol' TD so relaxed he was smoking a cigarette and enjoying the view.

That cigarette nearly undid Mr. Cool. To reach the cigarette pack kept in a sleeve pocket of his flight suit, Callahan had to undo the six-point shoulder harness — up through his legs, around his waist, and over his shoulders — that strapped him into the aircraft. Once he arrived over the carrier, all the procedures we had been practicing for weeks took over: the wheels were lowered, hook extended, canopy opened, power set, flaps set, but nowhere on that checklist did it say to reattach your shoulder harness.

TD said he came down the glide slope like on a rail. He caught the middle wire of five, which was perfect. The airplane came to a sudden stop, but ol' TD didn't. Fortunately, he had his helmet's visor lowered, so when his face smashed into the instrument panel, his eyes were protected. However, the lower edge of the visor cut a neat incision across each cheekbone and over the bridge of his nose. TD said later, "I sat there in the cockpit stunned, blood pooling in my lap, and thought, 'Geez, these carrier landings are tough! I don't know if I can make five more of these things.'"

TD figured out the problem, strapped himself into the airplane, and completed his qualification.

T-28 on short final to carrier deck

***

Air sickness is one thing in more than fifty years of flying that I have never experienced. Paul Burnham, one of my fellow preflight classmates, wasn't so lucky. And what a shame it was. Paul had dreamed of flying all his life. I imagined Paul as the kid in the front row of his first-grade class picture wearing the leather flying helmet. I bet Paul's bedroom had dozens of model airplanes hanging from his ceiling. Unfortunately, in flight training, the minute Paul left the ground, he became ill. Flight instructors became reluctant to take him up. Navy doctors tried various medications to settle his stomach, without success. Paul begged and pleaded to be able to continue in the program and was even allowed to make his first solo flight. But that was a mistake. Shortly after taking off, Paul radioed the tower. "Errr, Saufley Tower, Mentor Two One. Over."

"Mentor Two One, Tower, go ahead. Over."
"Tower, Two One, I have a problem. Over."
"State your problem, Two One. Over."
"I can't turn. Over"

"Say again, Two One, you can't turn? Over."

"Errr, that's affirmative. Over."

"Two One, Tower. Why can't you turn? Over."

"Because I throw up. Over."

Paul was heading west, well over Alabama, when two instructor's planes finally caught up with him. By talking Paul through a gentle 180-degree turn, the instructors got him lined up for a hundred-mile straight-in landing to NAS Saufley Field. It was reported that Paul's final landing as a Navy pilot was perfect, the main landing gear simply chirped as Paul smoothly settled onto the runway.

"Good for you, shipmate. Best of luck in the fleet as a swabby."

My friends all decided our favorite Paul Burnham quotation was, "I really love flying. It just makes me puke!"

One of the absolutes in military aviation is you would much rather die than be recognized for making a stupid mistake. A classic pilot error that happens with some frequency is forgetting to lower your landing gear and making an inadvertent wheels-up landing. Oh, the humiliation such a miscue will bring, if you survive.

Generally, such oversights happen during some other emergency, like an engine failure, or a fire, when the pilot becomes otherwise preoccupied. But sometimes it's just unexplained stupidity. A pilot of a multi-engine R-4D had survived a wheels-up crash and was being interrogated by the accident investigating board. "Lieutenant," they said, the tower reports it transmitted repeated warnings: 'Wheels not down! Wheels not down!' Did you not hear those warnings?"

"Yes, sir. I heard them."

"And what was your reaction, hearing those warnings?"

"Well, sir, I turned to my copilot and said, 'Listen to that. Some dumb bastard is going to make a wheels-up landing.'"

Inexperienced student pilots are prone to such accidents, and for that reason the Navy places an observer armed with radio communications and a Very pistol (signaling device) at the end of student-pilot practice fields, watching for such oversights.

A first-time solo pilot flying a T-28 trainer was making a landing at Baron Field in Pensacola when the observer noticed the plane did not have its wheels down when turning onto final. First, announcing, "Wave off, wave off. Wheels not down!" he then fired the Very pistol, sending a brilliant orange smoke trail into the sky before the approaching aircraft. Still the student continued, so focused on his line-up, airspeed, and altitude that he didn't see or hear the frantic warnings.

The plane touched down amid a shower of sparks, smoke, and flame and slid a hundred yards down the runway. Just as it appeared the aircraft might be slowing, a wingtip caught an obstruction at the runway's edge and flipped the aircraft over on its back.

The crash trucks, responding immediately, raced down the runway to the smoking airplane. Inside the cockpit the humiliated student, no doubt envisioning the hell he would face if he survived, had the presence of mind to remember the T-28 had a hand crank in the cockpit for lowering the wheels manually in the event of a hydraulic failure.

As the fire trucks approached the smoking hulk, to their amazement they saw two wheels slowly start to emerge from the underside of the aircraft's wings. Did that student really think he might get away with that?

A well-known warning about wheels-up landings is that if you have never made one, someday you would try. What about me? Did I ever come close? Yes, I did.

It happened while I went through the RAG (Replacement Air Group Training). Prior to reporting to my squadron after getting my wings, I was sent to the Naval Air Station Jacksonville, where I received flight training in the P2V7, the aircraft I would be flying in the squadron. The P2 is a big airplane, slightly larger than a WWII B-17. It is considered a four-engine aircraft but has the unusual configuration of two reciprocating propeller engines and two jet engines. The plane could fly safely even if one of the propeller engines failed, but landing in that configuration became rather tricky, since reversing the one engine is necessary in order to stop. Keeping the airplane on the runway takes a lot of one-sided footwork on the rudder and brakes. For that reason, single-engine landings were practiced on every training flight.

P2-V7 Neptune "Two Turning, Two Burning"

In a normal landing, wheels are lowered early in the approach, but because the landing gear adds excessive drag, when making a single-engine landing, the wheels are not lowered until after the landing is assured.

I was practicing a single-engine landing with the instructor, and the approach was going well. The jet engines were at idle, and I was adjusting the one-engine RPMs as well as applying asymmetrical rudder forces to keep the plane aligned with the runway. A warning device in almost all aircraft is a gear horn, which sounds if the wheels are not down whenever the throttle is reduced. The horn had been blaring during my approach, which was expected and ignored. Approaching short final and thinking primarily of the landing flare and the reverse procedure, it suddenly struck me the wheels were not down.

Applying full throttle to both jet engines and the recip, I loudly announced, "I'm waving this one off, sir. Taking it around."

"I think that's a damn good idea," the instructor said. Such an error would normally receive a poor flight evaluation. However, at the postflight briefing not a word was said. Had the instructor also been distracted and unaware? Or perhaps he just felt my lesson was learned. Whatever the reason, never in the next three thousand hours of flight did I ever come that close to a wheels-up landing again.

# # #

Chapter Ten

Rosey

I have a thirst for wandering across the United States for weeks at a time on my Goldwing, camping under the stars, and covering anywhere from 6,000 to 7,500 miles. It's an experience of discovery to set out in a general direction with perhaps only a few specific stops in mind, with new sights to see, people to meet, and stories to learn. The most enjoyable part of each day occurs at my campsite with my butt parked in a folding camp chair, computer in lap, as I document the day's discoveries.

Dodge City, Kansas, 2011

The challenge of motorcycle travel and camping is dealing with rain. I can handle cold, unimproved roads and heavy traffic, but rain is treacherous. It's not the slipperiness; it's a visibility issue. I look out though two sides of three different surfaces; my trifocal eyeglasses, the helmet's visor, and the motorcycle's windshield. Raindrops covering both sides of those translucent lenses make it like trying to see through Coke bottles.

The other, not-so-hazardous, but problematic concern is getting caught in a campground by a surprise thunderstorm, then having to bundle up one hundred pounds of wet tent, sleeping bag, clothes, cook gear, and equipment for the day's ride. When wet nights are forecast, I hunker down in a motel.

***

The campground in Wisconsin appeared inviting enough: wi-fi was provided, as were hot showers, and campsite electricity to charge my numerous devices. The forecast was for a clear, starlit sky. The nighttime weather prediction proved true, but the morning weather had other plans. When crawling out of my two-man tent — just big enough for me and my many bundles — sprinkles of rain greeted my emerging head. Uh-oh. Skip the hot shower. I needed to load up and get out of there.

I completed the normal forty-five-minute loading process in less than fifteen and headed toward Lake Superior, my targeted destination for the day. However, just a mile into the ride the skies darkened, and within minutes the light shower became a deluge. I couldn't see fifteen feet in front of me. I needed to stop and find shelter, but the forested topography of northern Wisconsin was shy of a roadside sanctuary.

I've learned some tricks when dealing with heavy rain. I lower the windshield, so to see over it — a nice feature of my late-model Goldwing. And I stand up on the foot pegs and let the wind blow the raindrops off my helmet's visor. I applied both of these methods but still rode nearly blind. Taking off my water-smeared glasses wouldn't help; I can't see a thing without my specs.

Fortunately, I came upon a general store with a canopy over its gas pumps. I didn't need gas but pulled in between the pumps, enough out of the way to not block the drive.

I trudged inside in rain jacket and rain pants, still wearing my dripping helmet, and sought out the proprietor to ask permission to leave the motorcycle where I parked it. The general store in this little hamlet was busy. In one corner sat twenty members of a coffee klatch on mismatched folding chairs, which I surmised they brought in with them. I guessed this was a regular morning gathering.

The proprietor turned out to be a proprietress, and her response to my dripping arrival and apologetic blocking of her gas pumps was, "Oh, pshaw, honey. Grab yourself a cup and sit a spell. This weather is bound to move on sooner or later." After I introduced myself to her, she announced to the group, "Hey, this is Jim. He's from Indiana and riding to Montana. Find him a chair and a cup."

These are the opportunities I look for: interesting people gathered in friendly conversation, telling their stories. One of my favorite authors, William Least Heat Moon, has written two wonderful books about travels like mine, *Blue Highways* and *River Horse*, in which he meets fascinating folks across North America and relates their tales.

After peeling off my wet gear, I answered the questions I am oft asked, such as "Where are you from?" "How long have you been out?" And "Why do you ride alone?" The riding-solo question is one everyone asks. Motorcycle travelers are mostly seen in groups. "Isn't it safer that way?"

Yes, I suppose it is. I carry a cell phone in case of mechanical trouble, but more importantly, riding alone gives me the spontaneity I feel I need. I have a rule when riding: if I fly by an interesting sight at seventy miles per hour, I'm to stop and go back to investigate. You can't do that when riding in a group. Just the other day provided a good example. I passed a yard filled with dozens of bizarre metal sculptures. I went back and found the artist, Greg Koeppel, a fifty-something University of Wisconsin art major, mowing the lawn. Greg worked at the local high school as an art teacher and loved art. Greg had built his own house, an igloo-shaped building covered with solar panels. "I'm off the grid," he announced, beaming. I spent forty-five minutes touring his creations, which included an eight-foot metal antelope, six-foot birds pulling steel cable worms out of the earth, and a ten-foot Christ on a cross. Greg then invited me to stay for lunch, "If you didn't mind veggie burgers and tofu," he said. Although it sounded tempting, I thanked him kindly and declined. The road was calling. I thoroughly enjoyed visiting with Greg. You can't do that when riding in a gang of motorcyclists.

The general store's coffee club was mostly summer people who have cottages in the area and come to Wisconsin to escape the summer heat of Chicago and other urban jungles. They have probably been doing this for years, starting each day with coffee and gossip, then fishing, boating, and napping until cocktails at five.

Ralph talked about a bear that broke into his garage and ate a bag of dog food. "I chased him away by blowing a whistle," he said, "but damnedest thing — next day I saw him hiding behind a tree across the street with another bear. It's almost like he was saying, 'Now, in that garage. That's where he keeps the good stuff.'"

All laugh. Then a half-dozen folks chime in with their bear stories.

Irene told one about raccoons in her attic, and that topped the bear story. She reminded everyone how she had set raccoon traps, but the clever rascals had removed the peanut butter bait without getting caught. Yesterday, she said, after taking a shower she thought she heard a noise on the porch. Wearing nothing but her brassiere and underpants, she went out to investigate. Not seeing anything on the porch, she went down the steps to the front stoop. Just then, around the corner of the house, came a big raccoon, apparently on his way to the porch latticework, to climb to his nest in the attic. Seeing Irene in his path, the raccoon stopped, reared back on his haunches, spread his front legs, and gave her a big hiss. "The audacity of the SOB," Irene said. "Not knowing what else to do, I raised myself up big and tall, spread my arms, and hissed the little bastard right back. That coon blinked a couple of times, lowered himself, and high-tailed it back around the house."

All agreed: Irene hissing in her bra and panties would be enough to make most folks run and hide.

I sat next to George, who remarked upon learning I lived in Bloomington, "Oh, my God. I got my masters at Indiana University." That was our first connection. George went on to tell of becoming a Cub Scout as an eight-year-old and having a sixty-two-year relationship with scouting.

"So you became a Scout executive?"

"Yep. Great career. I loved it."

"Well, I have three sons, two of them Eagle Scouts. And I was in scouting for a number of years myself, from Cubs to Sea Explorers."

That couldn't have pleased George more. "In fact," I said, "one of my Eagle sons is running across Wisconsin and Michigan this summer, and I'm going to rendezvous with him three weeks from now."

"Running across Wisconsin and Michigan?"

"Yep. Brian is a long-distance runner and in 1998 spent nine months running across the United States. He now has a goal to run across every state by the time he is fifty."

"I bet he is able to do that because of scouting," George said.

"Well, maybe," I replied. "His oldest brother, also an Eagle Scout, is conductor of the Indianapolis Symphonic Choir, a Butler University professor, a pilot, a runner, and a world traveler, and has performed in China, conducted at Carnegie Hall and the Kennedy Center, and teaches in Japan and South America."

"I bet he is able to do that because of scouting."

"I'm heading to a campground in Glacier National Park and then returning here to meet Brian," I told George.

George suggested that I also could do those things because of scouting. I debated telling him that the only things I remembered about scouting were learning how to tie a square knot and to smoke cigarettes, but I decided against it.

Soon the rain lightened to a sprinkle, and I bid my cordial hosts good-bye, mounted Rosey, and headed up the road.

I marvel at the sights I see through my Goldwing's windshield. There is no mode of travel that puts you more into the surrounding scenery than travel on two wheels. The settings stimulate all senses; you smell it, hear it, and vibrate with every change in the road's surface. With your feet a mere six inches above the passing hardtop, you delight in watching your shadow bounce and hop over the road's shoulder and guard rails. No sunrise or sunset ever wrapped itself around you so completely as one seen from the seat of a bike. Cresting a hill and suddenly having a sight like the Grand Teton Mountains fill a 180-degree field of view prompts an admiring expletive every time. I guarantee it. I have seen sights like that over and over, from northeast coastal seascapes, to southern bayous, to western plains.

Grand Tetons, 2011

***

Winnett is a dusty crossroads town on the plains of eastern Montana, possessing just a few buildings. It's actually not at a crossroads at all but sits a mile south of the main highway. My map showed no other towns within thirty miles, which concerned me, because I had taken a chance that Winnett would have a gas station; my near-empty gas tank would not make it another thirty miles. To my relief, the dozen buildings in Winnett included one with a gas pump in its side yard.

One of the other buildings appeared to be a former filling station, but now a Coca-Cola sign identified it as the Kozy Kitchen Café. It was past noon, I hadn't eaten since my SlimFast breakfast drink that morning, and I was hungry. Parking Rosey, I saw one woman on a ladder painting the exterior of the building, and another puffing a cigarette under what must have been the former gas station's canopy. The smoker didn't ask if I wanted to eat; she just said, "Follow me," and led me through a door, down a hall, and into a smallish room with a few tables, one occupied by four middle-aged women who had just finished eating and were now smoking and talking.

I ordered a fried egg sandwich and a cup of coffee.

On the wall, above my table was a reprint of a 2005 *Gourmet Magazine* article about Kozy Korner Café. "What!" I asked the woman who took my order. "Is that for real?"

"Oh, yeah," she said. "That's before my time here, but it's a real funny article, especially about Buck."

The writers were the well-known *Gourmet Magazine* foodies Jane and Michael Stern, who had been doing a series of articles about great restaurants across the U.S. but had not found one in Montana. A friend told them to check out Kozy Korner in Winnett. "They have the best pancakes on earth."

Jane and Michael arrived in Winnett early one morning, expecting to see a parking lot filled with pickup trucks owned by ranchers there to enjoy the best pancakes on earth. Instead the lot was empty. The time was 6:50. They waited a few minutes, wondering what to do. Finally the door opened, and a grizzled old codger, without a word of greeting, signaled them to come inside. He was Buck Wood, the owner. He and his wife, Ellen, had been tipped off that the *Gourmet* writers were coming. Ellen told Buck, "Now you be nice to these people."

The authors, used to elegant dining, described the place as having all the panache of a truck stop or a county jail. Jane asked Buck if she could have some water, and Buck answered in his grumpiest manner, "There's water in your coffee." He then brought two ice-filled glasses of water. Jane then asked for milk for her coffee. "What does this look like, a dairy?" Buck replied. He took Jane by the hand, led her over to the refrigerator, and poured milk from a carton into Jane's cup. All the while, he was giving Michael a wink on the sly. The Sterns had pancakes and loved them.

My egg sandwich was good, but I should have had pancakes.

***

In 2009 I rode on what I called a "Down East" adventure. Back in the sailing days, ships leaving Boston Harbor used to sail northeast up the coast to Maine before catching the trade winds and venturing across the Atlantic. The prevailing winds to Maine provide a downwind sail, and thus the expression "down east." Leaving Indiana, I traveled through Pennsylvania, New York, and Connecticut, and then on up to Maine, but stopped to visit the Gettysburg National Battlefield and later camped in Connecticut near Colchester, where I was doing research for a book I was writing, *Great Lakes Skipper*. Out of Bar Harbor, Maine, I took the Cat, a ferry to Halifax, Nova Scotia. The Cat Ferry is a large, high-speed catamaran capable of carrying 250 cars, 725 passengers, and fourteen motor homes or tour buses. The crew had the vehicles line up by type to be loaded in appropriate sections of the ship. There were twenty motorcycles, and I thought it was interesting that a rider on a very tall unicycle was leading our group. I can't explain that one.

We parked in a special section of the car deck and were told to put the bikes on their side stands, rather than use the center-stand. Although we were advised we could safely leave them that way, I noticed tie-down chains hanging from the ship's bulkhead, so as an extra precaution I used the chains to lock Rosey in place.

The Cat's two-hulled design lifted the ship above the waves and normally provided a very smooth passage. That day, however, the seas were unusually turbulent, and regular commuters said they had never seen it so rough. The ship had cocktail bars, casinos, and lounges, but walking about became difficult, and most passengers were scrunched down in deck chairs, looking very green.

I felt reassured having chained my bike in place, but now I worried about all those other bikes sliding around the deck below. We weren't allowed to go below, so I hoped for the best. Turned out that, although there was noticeable rearranging, Rosey showed no signs of intimate caresses from any of the other motorcycles.

My Nova Scotia plan was to have my wife fly into Halifax and spend a week with me for rides around that enchanting peninsula. Michele is not into camping. She needs flush toilets, electric outlets for her hairdryer, and a cosmetic mirror for her morning's preparations. No matter that her leather garb and full-face helmet were going to conceal all those laborious rituals; she just has to look good for the departure. And she always does.

One of my high school classmates, Bill Robinson, has a sister-in-law living in a condominium in Halifax. Dorothy graciously arranged for us to rent a guest suite in her condo and helped plan our itinerary while visiting her country. And what a wonderful hostess she was. When we arrived, the refrigerator was stocked with various bottles of wine, beer, and snack foods. Every night after we returned from touring, there would be a note under the door with suggestions for the next day's sightseeing.

During my first night in Halifax, Dorothy invited me for dinner. Michele wouldn't be arriving until the following day. Dorothy also invited her friend, Stan, a former airline pilot, to join us. We talked about various places they thought Michele and I ought to visit. Peggy's Cove would be number one on the list.

Described as the most-photographed spot in all of Nova Scotia, Peggy's Cove has it all: lighthouse, rocky coast, gift shops, lobster boats, colorful cottages, and crashing surf. Tragedy had also visited. On December 10, 1999, Swiss Air Flight 1800, out of New York, started its trans-Atlantic flight, caught fire, and plunged into the sea near the beaches of Bradford and Peggy's Cove. Two hundred and twenty-nine passengers and crew were instantly killed. Operations to recover the bodies and wreckage from the ocean floor were set up in both of those communities.

Stan knew a lot about the circumstances of that crash. The fire apparently started in the entertainment console aboard the aircraft. The pilot-in-command (PIC) was a standardization pilot, put aboard this flight to give the pilot in the right seat his periodic check-ride. The PIC was a by-the-book guy. When smoke first appeared in the cockpit, the PIC pulled out the emergency checklist and started going through the procedures. The voice recorder revealed that the fire spread rapidly, and the copilot was heard to say, "We have to get this plane on the ground *now*!" But the by-the-book guy said that the plane was too heavy and that they needed to dump fuel and complete the checklist.

The aircraft made two passes over the Halifax airport. After the second pass, the plane suddenly nosed over and dove straight into the ocean. All souls aboard were torn to shreds and died instantly. It was ruled pilot error even though the pilot-in-command did everything according to procedures. Had he landed overweight, it might have collapsed the landing gear or even torn off the wings, but there would have been survivors. Stan said that during his last visit to the Bradford Memorial site, a couple and their two young children were also visiting the site. The father pointed to two names on the memorial and said to the children, "Those are your grandparents."

Stan said, "That's when I lost it."

Peggy's Cove has a lighthouse sitting on its granite shore. The lighthouse also serves as a post office, and postcards bought there can be mailed from that location. I wrote a postcard to my ninety-year-old mother but, before dropping it in the mail slot, had Michele take a picture of me holding the card. I gave my mother that picture later when next I saw her.

Peggy's Cove, Nova Scotia, 2007

Several gift shops sit on the roadside of the village of Peggy's Cove. After visiting a number of them, we found a lobster restaurant and had two broiled lobsters while sitting on the deck under the brilliant blue sky, listening to the squawk of seagulls. It doesn't get much better than that.

Our next day's ride took us to the little town of Mahone. It had a museum, and Michele, a native-born Frenchie, understood that the French had settled Nova Scotia. The museum had a list of the early family names in Mahone, and they were all German. What was that about?

An excellent museum docent gave us the sad story about the French in Nova Scotia. They did indeed settle this land as early as the 1600s. For a hundred years they farmed the nutrient-rich floodplains, primarily near Grand-Pre, and devised a system of dikes to control the flooding so their crops were not washed away during the high tides. The French called themselves Acadians.

During this time the British were exploring and claiming foreign lands and were fighting the French in Europe. The Brits saw Nova Scotia as a key to their domination of the Atlantic and considered the Acadians French, regardless of what they called themselves. The British expelled the Acadians from Nova Scotia in 1755 in an action referred to as Le Grand Dérangement. Thousands of Acadians were put on ships and deported to the American colonies and several countries in Europe. Their homes and farms were burned, so that they would not return. Tragically, because of their Catholic faith the Acadians were rarely accepted wherever they went. The king of Spain, however, allowed the Acadians to settle in the Spanish-held territory in Louisiana. Today, many of the Cajun population of New Orleans can claim Canadian origin.

A rainy day in Halifax gave us the opportunity to take a Gray Line tour of the city and learn about two significant events in Halifax history. The first was the explosion of a heavily laden ammunitions ship, the *Mont Blanc*, which collided with the steamer *Imo* in Halifax Harbor December, 1917, during World War I. The *Mont Blanc*, loaded with TNT, gasoline, and armament, exploded with the force of the two bombs dropped on Nagasaki and Hiroshima combined. All of the water was blown out of the harbor. Two thousand people were killed. The city for miles around was flattened. And then, as if the explosion weren't bad enough, the worst blizzard ever experienced by Halifax descended on the city that very night. Those not killed by the blast were found the next day frozen to death.

A tour of the Halifax Museum revealed artifacts of the sinking of the *Titanic* in 1912. Halifax was the primary base of operations for the recovery activities. Utility boats of all description were pressed into service to go out and pick up floating bodies. Two hundred twenty-nine bodies that were not claimed by relatives were buried in a Halifax cemetery. Many of the rescuers picked up floating articles of clothing or pieces of the ship they kept as souvenirs. Many of those items — a child's pair of shoes, a hairbrush, a deck chair — have now been turned over by the families to the museum.

In the movie *Titanic*, the leading character, played by Leonardo DiCaprio, was named Jack Dawson. The character was fictitious, but the director used the name of a real victim, J. Dawson, who worked in the ship's boiler room. Following the release of the movie, young female fans flocked to the cemetery, adorning J. Dawson's gravestone with hundreds of Teddy bears and flowers. Boiler-tender Dawson would have been pleased.

Our last day of motorcycle touring was to the Bay of Fundy, to see its forty-foot tides. The bay is a narrowing stretch of water between Nova Scotia and New Brunswick. Tides push into the bay with great force, resulting in the dramatic change in the water level every six hours. We took pictures of fishing boats lying on the mud of inlets only later to be floating forty feet higher, tied to the pilings.

Bay of Fundy, Nova Scotia, 2007

Nova Scotia is a beautiful province, urging us to consider a return visit someday before too long.

# # #

Chapter Eleven

"Yellow Blood, Yellow Blood!"

Our Navy antisubmarine warfare (ASW) squadron flew twelve- to sixteen-hour patrols searching for Soviet submarines. In addition to the hot war going on in Vietnam, the 1960s Cold War with the Soviets had American schoolchildren "ducking and covering" under desks in schools across the U.S., preparing for a nuclear weapons attack. The world situation was tense.

I served on active duty with the Navy from June 1962 until July 1967. Flight training at bases in Pensacola, Corpus Christi, Norfolk, and Jacksonville consumed the first year and a half of that hitch, followed by three and half years of duty while assigned to Patrol Squadron Twenty-One, home-based in Brunswick, Maine.

No event in recent history brought the United States closer to World War III than an incident that occurred in October 1962. U.S. and Soviet relations at the time were bellicose. Eighteen months earlier, President Kennedy attempted to overthrow Fidel Castro, the Communist dictator in Cuba, with the disastrous Bay of Pigs invasion. The Soviet premier, Nikita Khrushchev, in support of Castro, brought ballistic missiles to the Cuban island, putting nuclear warheads ninety miles from U.S. shores. The event became known as the Cuban Missile Crisis.

President Kennedy went toe-to-toe with Khrushchev and demanded he remove the missiles in exchange for U.S. concessions in Turkey, or American troops would invade Cuba and remove them for him. Khrushchev acquiesced, the missiles were removed, and Americans breathed a sigh of relief, assuming the crisis was over. Not so!

Although the missiles were gone, the Soviets maintained a large contingent of ships, aircraft, and armed troops on the

Communist island. The American military watched Cuba closely. Continuous Navy patrols were flown around the island. The Air Force had jet Phantoms in Key West with engines running twenty-four hours a day, ready to launch to meet any hostile activity. Both sides had nervous fingers on the triggers. The tension continued for many months after the October 1962 crisis.

I checked into my squadron in Maine in January 1964 and received an immediate assignment to a crew being sent to Key West as part of a detachment flying patrols around Cuba. Our squadron had fourteen combat air crews flying P2V7—Neptune aircraft. A crew included eight enlisted crew members and four officers. Three of the officers were pilots: plane commander, copilot, and navigator. The fourth officer served as the tactical coordinator.

I was assigned to the executive officer's crew. The XO didn't fly often because of his other squadron responsibilities. That meant his copilot conducted many of the flights as acting patrol plane commander (PPC), and I, as the third pilot, moved up to the copilot seat. As a new officer in the squadron, hungry for flight time and more responsibility, I found that to be a good deal.

The Key West operation was serious business. Briefings before every flight were conducted in a top-secret bunker illuminated with red lights. State-of-the-art radar screens surrounded the briefing room. A joke suggested that if someone scratched his butt in Cuba, it could be seen on radar in Key West. Large maps of Cuba hung from the walls. Plain-clothed civilians with top secret badges moved about in the shadows. Patrolling aircraft received "Yellow Blood" messages transmitted every five minutes. The messages were meaningless; "Yellow Blood, Yellow Blood, Tango-India-Mike-Sierra" meant nothing. However, every day, three new four-letter code groups did mean something. One group meant the plane's navigation was screwed up, get back on track. A second group meant there was an unidentified target in the area; be alert. And a third group meant there was an enemy aircraft in the area and you appeared to be under attack!

The seriousness of the situation in Cuba was brought home to me when our commanding officer told all air crews, "If something catastrophic happens to your aircraft while on patrol [he didn't define "catastrophic," but we understood that to mean the enemy's involvement], head for open water. Do not crash on Cuban soil." Six months before, Russia shot down a Central Intelligence Agency's U-2 aircraft, flown by Francis Gary Powers, and negotiations for Powers' release became national headlines.

Rigging ships during Cuban Missile Crisis

Our first couple of patrols were routine. Flying at just three hundred feet above the water, we spent the hours investigating and photographing the decks of cargo ships – called "ship rigging" - sailing into and out of Cuban ports. On our third patrol, after six hours, the plane commander announced he was going to take a break, and I slid into the left seat as pilot in command. The PPC had no sooner dozed off in the back of the plane, when I came upon three Soviet destroyers and eight Soviet PT boats, a sight we didn't see every day!

Soviet destroyer

My wet-behind-the-ears inexperience saw this as an opportunity for heroism, so, with visions of air medals dancing in my head, I told the crew to get ready for some momentous photographs. I descended to masthead height and flew across the decks of the ships, so close we could see the open-mouthed expression on the faces of the Soviet sailors. I made four passes, all the while imagining what I would say to President Kennedy when he invited us to the White House to offer his congratulations.

The swooping and diving of our aircraft awakened the plane commander. Charging to the cockpit, he demanded to know what the hell was going on. Despite my excited explanation of our heroic discovery, the PPC asked, "Did you send an alert to Key West intelligence? Did you report the sighting in an encrypted message to squadron communications? DIDN'T YOU EVEN THINK TO WAKE ME UP?"

"Errr, no sir. I guess I didn't."

Just then the radio operator came over the intercom, "Flight, Radio. We just got a Yellow Blood message, Kilo-Yankee-Alpha-Gulf. Sir, they say we're under attack by enemy aircraft!"

Looking out the starboard windscreen, we could see three Russian MIG-21 fighters, five thousand feet above, in classic attack formation, poised to dive down and do some bad things to our airplane.

Russian MIG-21 fighter

We were a thousand feet above the water. With no time to switch seats with the PPC, I immediately rolled the airplane onto its port wing and dove for the surface of the Gulf. I thought, When those MIGs pounce, they're not going pass below us. I remember those moments as if they were yesterday. Skimming along the surface, spray kicking up from our prop tips, I was hunched over the yoke with both shoulders up beside my ears like a kid in a snowball fight, waiting for something to slam into the back of my head.

The measure of time was impossible to estimate. It felt like a lifetime but may have only been minutes from the time we received the Yellow Blood message until the cavalry arrived. Two Air Force F-4 Phantom jets came screaming past us from the opposite direction. Not much more than a blur, they disappeared briefly, then reappeared, joining up on each wing tip. The pilot on our starboard side looked over and gave us a smart salute. His gesture said it all: "The playing field has just been leveled."

Air Force F-4 Phantom

The MIGs never left their perch. Perhaps it was my rapid escape that caused them to hesitate, but most likely it was the arrival of those Phantoms that sealed the deal to stand down. The Phantoms escorted us back to Key West.

The plane commander covered for me at the debriefing, claiming the discovery of the Soviet ships and photo taking happened so quickly we didn't have a chance to report the sighting before the MIGs appeared.

We didn't get any air medals, and we didn't get invited to the White House, but we did have a good story to tell at the Officer's Club that night.

***

A postscript to that story occurred fifty years later. In January 2015 I spoke to a Military Officers of America Association in Lady Lake, Florida, about a book I had written, *Two Turning, Two Burning, Memoir of a Naval Aviator*. Although most of the attendees were husbands and wives, one single, rather attractive woman was introduced as the widow of an Air Force general. She sat near the front of the room, and I couldn't help but notice she was really enjoying my presentation. When I started telling the story of the Cuban incident, I saw she began silently clapping her hands.

After the presentation, people came forward to buy signed copies of my book; first in line, she wanted three copies. I told her I noticed she seemed particularly interested in the Cuban story.

"Yes," she said. "My husband was flying one of those Phantoms in Key West."

# # #

Chapter Twelve

It Just Doesn't Get Any Better than This

Taking command of a forty-foot sailboat in unfamiliar waters with twenty-knot winds blowing is a challenge. Nevertheless, having sailed for years on smaller boats emboldened me to think I could handle it. When I contacted the Moorings Charter Company in the British Virgin Islands in 1996, they confirmed my qualifications after evaluating my sailing résumé and agreed to turn over one of their $300,000 yachts to me as skipper-in-command.

The crew would be my wife, Michele, and a fellow employee, Doug, and his wife, Carol. I had sold Doug a small sailboat I owned in Indiana, so Doug had basic sailing skills. The four of us had planned the eight-day cruise in the BVI a year earlier and excitedly looked forward to the adventure.

The Moorings base is in Road Town, Tortola, BVI. Several other charter companies operate out of Tortola, but after careful research Moorings appeared to be the largest and the one with the most complete services. On our arrival, a porter met us at the airport carrying a sign, "Welcome Jim Stark." He gathered our luggage and handled transportation to the Moorings' base.

We would be sailing a forty-foot, Beneteau-built sloop, named *Sailbad the Sinner*. We hoped its name wouldn't be an omen of our next eight days. Moorings' services included stocking our galley with the breakfast, lunch, and evening meals we had ordered, as well as beer, wine, and liquor. Half our evenings would be spent at island restaurants. The other meals, of chicken, fish, shish-kebabs, and steaks, would be cooked on a grill attached to the boat's stern railing.

The day before our departure, Moorings conducted thorough briefings regarding customs, restricted areas, chart reading, and how to contact Moorings for assistance if needed while under way. A lovely young lady came aboard and spent an hour going over all the boat's features and operating systems. In addition, first-time charterers were given a "friendly skipper," who sailed out with us the first day and was with us for three hours, making sure we were comfortable with the boat before he took a water taxi back to the base.

Winds were robust that first day, blowing a steady twenty knots. Doug, the friendly skipper, and I put a double reef in the sail to help control the boat. Beating into the wind meant the boat was heeled over thirty degrees, with oncoming four-foot waves breaking against the bow and washing down the deck. Our seven-knot speed added to the twenty-knot wind howling in our ears, as well as the salt spray in our faces, made for an exciting introduction to the BVI.

Our destination the first day, Peter Island, required us to maintain our heading into the wind. Downwind sailing, even in similar blustery conditions, is much more relaxed. Moving with the breezes allows the boat to sail on an even keel and eliminates the gusts in your face and the breaking waves against the bow.

On arrival we tied to one of the mooring buoys in the bay and, after taking a deep breath, relaxed, enjoying the accommodations of our boat. Down below were three generous staterooms and two heads, each with showers. The galley included a refrigerator, a freezer, and a stove-top cooking surface. The saloon had comfortable seating for six, plus a fold-down table if meals were to be eaten down below. It's warm in the tropics, ninety degrees, so almost all our shipboard meals would be taken on deck in the large cockpit, under the sun-shading bimini top. The cockpit also had a table to sit around.

A boat the size of *Sailbad* has numerous complex electrical and mechanical systems, systems, we learned, that had occasional glitches. One of the advantages of having another man aboard is being able to muscle the stubborn problems into submission. Particularly when at the same time needing a steady hand at the helm.

The following day we were again headed into the wind toward Cooper Island. It is surrounded by reefs, and we had been briefed that there was only one safe entrance into its bay. After getting lined up, we prepared to start the engine, lower the sails, and motor into the bay. The forward sail was a furling jib. When operating properly, it was a simple matter to pull the jib's lanyard and have the sail retract and become secured around the forward stay; however, it jammed. The line that recoils the sail had jumped off its drum and became entangled. With the sail flapping and snapping in the wind, I steered away from the bay's entrance while Doug crawled out on the narrow bow to wrestle with the knotted line so we could retract the jib. At last successful, we motored into Cooper Island. Regrettably, there were two other occasions during the next few days that we had to deal with the same furling-jib problem.

Then there was that fussy outboard motor on the boat's dinghy. It occasionally coughed and sputtered and quit running. We had been successful in restarting the motor in those instances, until the night we took the dinghy into shore at Norman Island for dinner. It quit just as we arrived at the restaurant's dock. Knowing we would need transportation back to the boat after dinner, Doug and I worked on the motor for several minutes, but without luck. We decided to eat and worry about the motor later. After dinner, with the night now dark, we asked another patron to take the girls back to our boat, and Doug and I would row back. It shouldn't take more than twenty minutes, we estimated.

We rowed in a direction we thought correct, but we could not identify our boat among the dozens of other dark shapes in the bay. After thirty-five minutes of paddling about, Doug caught sight of the beam of a flashlight across the bay being waved frantically back and forth. Good work, girls. Guided by the light, we rowed on home.

We had been told our boat had a freshwater tank of 150 gallons, and we needn't worry about running out of water, even with the four of us taking showers. After two days we ran out of water. We called the Moorings' base to discuss the issue, and they also were surprised but directed us to a nearby marina to refill the tank. The tank held only 85 gallons, not the 150 we had been told. We had to make one other water stop before the end of our cruise, and that stop became complicated by the marina's loss of electricity, which prevented them from pumping water. Fortunately, electricity was restored a short time later.

The inboard diesel on *Sailbad* was an all-important part of our boat. Maneuvering to set the anchor or to tie up at a buoy, or getting into a marina or bay, would be extremely difficult without engine power. We happened to be heading to Cane Garden Island, made famous as the hangout of Jimmy Buffett, when we discovered the diesel wouldn't start. It wouldn't crank over, and the entire boat seemed to have lost all electrical power. It was a serious situation. Just getting someplace for help would be difficult without the engine.

I dug through the boat's trouble-shooting manual. It asked a long list of "yes and no" questions, all "no's" being answered by "Call Moorings." Before doing that, however, I crawled around in the engine compartment and discovered that *Sailbad* had three different battery systems. Each separate system was selected by moving a "T-handle" to the one chosen. Somehow the T-handle had slipped into a neutral position, and by turning the handle our problem was solved.

Lest you think these issues clouded our cruise, let me assure you the beautiful tropical waters, the gorgeous sunsets, the great snorkeling, and the exhilarating sailing trumped those problems by a mile.

The Baths on Virgin Gorda are an incredible phenomenon. Huge boulders, some as large as houses, lie on the white sandy beaches of the island. Piled on top of one another, they form rooms and chambers illuminated by shafts of sunlight. The pools of water in these scenic grottoes are warmed by the sun, thus giving the attraction the name Baths. The granite formations came from deep in the earth, formed by molten magma during prehistoric volcanic activity. These boulders forced to the surface, erosion over eons of time rounded them into the shapes we see today.

The Baths are a popular attraction, and guidebooks advised to arrive early to secure one of the mooring buoys. We did as suggested and the four of us swam ashore to marvel at the uniqueness of the site. It was fascinating to crawl through the caves, rooms, and crevasses. After an hour, the girls returned to the boat, while Doug and I continued to explore. I spotted an underwater opening fifteen feet from the beach and swam over to look down on it through my swim mask. A cave opening, no more than ten feet below the surface, seemed to be illuminated from within. I guessed there must be another opening to the surface in the cave and swam down to check it out. When entering the cave, I saw that the light came from a source thirty feet away, deeper into the cave. I had taken a deep gulp of air when diving down, so ventured closer to the shaft of light. The opening wasn't large but no doubt led to the surface at a place other than from where I had entered. When I stuck my head through the opening, I realized it would be a tight squeeze but figured if I could get my shoulders through the hole, the rest of my body would fit as well.

It was tight, but I got through, leaving a scrape or two of skin behind on the rocks. Doug had been treading water above when I dove down into the cave. Minutes later, when he saw me standing on the beach at the water's edge, he exclaimed, "How did you get there?"

Baths, BVI, 1996

The good service provided by the Moorings was demonstrated when we arrived at the Bitter End Yacht Club in the bay of Virgin Gorda. After securing to a mooring buoy, a Moorings' dinghy pulled alongside, asking if there were any problems that needed fixing. Fantastic!

"Yes," we replied. "There's a broken hinge on one of the cabinets, our dinghy's oars were stolen in Trellis Bay, the outboard motor isn't working right, and the jib furling system is all screwed up."

"No problem," was the reply. "We'll be back in a jiffy with tools to correct the problems."

How great. Turning over our boat's issues to the Moorings' service crew, we hitched a ride from a fellow yachtsman and went ashore to explore Virgin Gorda. A three-quarter-acre island in the bay, named Saba Rock, was owned by Bert Kilbride. I had read about Bert in my research prior to our trip to the BVI.

Kilbride, famous among diving circles, was a marine archeologist. Born in New England, he claimed to be a diver from the day he was born. He moved to the clear waters of the British Virgin Islands in the early 1950s, near Anegada Island, to search for the 228 known wrecks left behind by Blackbeard, Henry Morgan, and Sir Francis Drake. Bert discovered 138 of them.

His most significant discovery was the *Rhone* in 1958. The ship went down in 1867, and Kilbride brought enough artifacts to the surface, including the skull of the ship's carpenter, to fill a museum.

Bert started a dive company, taking customers down on the *Rhone*. His patrons included the rich and famous, such as Robert De Niro, Walter Cronkite, Geraldo, Jackie Onassis, and the eighty-year-old Dr. Benjamin Spock. In 1989 Kilbride purchased Saba Rock and opened the Pirates Pub. Doug, Carol, Michele, and I were enjoying Painkillers on the porch of Pirates Pub while I told them about Bert. It was then I noticed an old codger a few tables away, needing only an eyepatch and a parrot squawking obscenities to be an old pirate. "That's Bert Kilbride. I'd bet a million dollars."

I was right. Bert, eighty-four at the time, has been married five times. His current wife was thirty-four years younger. Bert once attributed his longevity to his love of vodka and his love of young women.

After introducing ourselves, we joined him at his table and heard some wonderful stories about his adventures. Bert discovered an anchor dropped on the bottom by Christopher Columbus. "I tried to get permission to bring it up to be displayed in a museum, but the Virgin Island authorities insisted it be left in the sea. No problem," said Bert. "Instead, I make thousands of dollars taking divers down to see the anchor."

Departing each morning to sail to a new destination involved a number of important preparations. I was surprised the Moorings had not published a presail checklist. The crew of *Sailbad the Sinner* created its own and later submitted it to the Moorings for their consideration.

It listed: check heat exchanger, check oil, clear lifelines, take down wind scoops, close valves, pump toilets dry, pump shower bilges, stow fenders, lock down oven, pull up swim ladder, close swim platform and dog hatches, lock drawers, close reefing cringle.

Any item overlooked could result in a mess to clean up or damp conditions with which to deal.

After six days of sailing, our crew had become experienced old salts. The gals as well as the guys could handle the sails and deal with capricious winds. After rounding the north side of Tortola Island, we were once again beating into the wind while heading for Norman Island, known as the setting for Robert Louis Stevenson's *Treasure Island*.

We were entering a narrow passage, and most conservative sailors would opt to drop their sails and run under their iron jenny (diesel engine), rather than have to tack back and forth in those waters. Experienced as we were at this point, we chose to sail through the channel just for the fun of it, and also because two other boats, a cat-rigged sailboat and a sloop, were also under sail slightly ahead of us. We wanted to see if we could outrace them. Whenever two or more sailboats are heading the same direction, admit it or not, it becomes an undeclared sailboat race.

Our crew's coordination was a thing of beauty. Carol handled the starboard jib sheet, while Michele was on the port side. Doug managed the main sheet, and I was at the helm. We made eighteen tacks during two hours of transiting the channel. "Stand by to come about! Ready about! Helms alee!"

Our catboat competitor was able to sail closer to the wind but wasn't as fast. The sloop was just sloppy, sails frequently flailing about after their tacks until brought under control. The race may have been undeclared but was nonetheless hotly contested. And the winner? No doubt about it.

Sailbad the Sinner!

Norman Island is the westernmost of the islands that circle Tortola and make up the British Virgin Island group. It is open to a wide expanse of sea to the west, and any storms rolling in from that direction hit Norman full force. As a result, the cliffs on that side of the island have been honeycombed with hollows and caves carved by winds and waves. Most are shallow, but six caves reached back into the rock fifty to one hundred yards. Scuba divers and snorkelers made the caves popular to explore.

After anchoring our boat in the Bight of Norman Island, we jumped into our dinghy with its now repaired outboard motor and rode to the cave entrances twenty-five minutes away. After tying our dinghy to a mooring buoy, the four of us donned swim masks, snorkels, and fins and swam to the rocky shore.

The caves, from the outside looking in, appeared dark and foreboding. Michele is not a strong swimmer and consequently not overly comfortable in the water. I wondered if she was going to pass on the cave exploration, but she surprised me, just as she did when first trying snorkeling in Cancun a few years earlier. She was hesitant that time, but once she saw the beauty of the underwater world, I had a hard time getting her out of the water. Yes, she wanted to go into the caves.

The Caves, Norman Island, BVI, 1996

    Michele clung to my T-shirt as we paddled into the abyss. I must admit it was creepy. We kept looking back toward the opening for the reassurance of daylight several yards away. Our imagination suggested slimy, scaly things rubbing against our legs. As mysterious and eerie as the experience was, its uniqueness should put it on everyone's list of places to see in the BVI, second only perhaps to the Baths.

    At last our adventurous curiosity was satisfied, and we returned to the dinghy. Much to our surprise, in our hour's absence a squall was approaching, and we knew we'd best get our little boat back to the big boat. We no sooner started the motor than a driving rain hit us, accompanied by strong winds. The blow was coming precisely out of the direction we wanted to go. Waves built quickly, splashing over the bow of the dinghy. The rain pelted our faces with such stinging velocity we couldn't keep our eyes open. Carol put on her swim mask, an obvious solution to the situation.

    Then Michele started to giggle, which she often does in threatening situations, but her laughter was so contagious that soon all of us were practically doubled over with hilarity. Not so out of control, however, that we didn't realize we had to bail out the boat. We scooped the water with our hands but barely kept up with the waves splashing over the bow. Then the motor quit. Oh, no. Not again!

We needed propulsion or those winds were going to push us out of the BVI, past St. Johns, and out to sea. Still the gales of laughter continued. Even more so when Michele admitted she'd just wet herself. But then, who could tell? Doug finally got the motor running and got us back to *Sailbad*. The rain stopped just as we arrived. Five minutes later, the sun burst through the clouds, erasing the gray, and left a brilliant blue canopy above.

After changing into dry clothes and assembling on deck for our sundowners, we repeated the refrain we had been singing all week, over and over: "It just doesn't get any better than this."

What a fabulous eight days we had spent. What sights, what thrills, and what fun. And how great it was to spend the time with good friends.

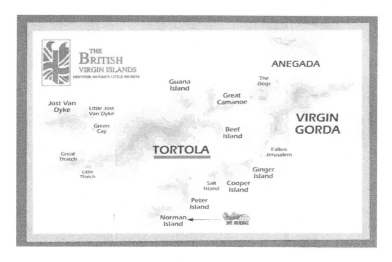

#

Chapter Thirteen

Sedona

Mishaps with motorcycles and police involvement do not always involve accidents. In 2003 I rode out west to tour Arizona and New Mexico. Prior to the trip, I visited with my friend Winston, who grew up in Oklahoma, to get tips on places to visit. Winston said when I went through Erick, Oklahoma, to give it a salute, because that's where he was born. When I found myself in Erick, I decided to take a picture of the town marker and get some gas.

I had been wearing my jacket that morning because of the cool temperatures, but now it was in the seventies. At the gas pump, I pulled off the jacket and draped it over the back of the bike as I filled the tank.

Once finished, I rode off, continuing my trip. The breezes felt cool as I rode, and that reminded me of the jacket. Straining to look back, I saw it was no longer where I had left it. I stopped beside the road and wondered how far it would be to the next crossover on the divided highway, and I hoped I could see where the jacket had blown off.

Then a highway patrolman, lights blazing, screeched to a halt behind me. I approached his car with some trepidation and asked, "Did I do something wrong?"

"Just lose something, did you?" he smiled.

He had parked at the filling station where I gassed up and saw my jacket fly off as I left. How lucky is that?

Then he said, "It took me some time to catch you. You move right along, don't you?

"Well, not too fast I hope ... thanks for the coat. I really appreciate it."

"You're welcome. Now ride safe, ya hear."

***

I don't claim to have special powers regarding the luck I enjoy, nor do I flaunt it. I've never purchased a lottery ticket. Certainly, I've had my share of flat tires, sick days, and bad investments, but invariably, whenever black clouds appear, they always seem to be followed by offsetting good fortune. Like the time I ran out of gas because of a faulty fuel gauge but was able to coast into a gas station. Or the time a wheel bearing overheated on a trailer I was towing, but I noticed the smoke rising from the hub when we stopped for coffee. Would you believe I had stopped in front of a truck garage that was able to replace the wheel bearing without even moving the trailer?

Sedona, Arizona, is said to be especially friendly to those seeking supernatural powers. I spent several days in Sedona in 2005 on a motorcycle trip, hiking its trails and marveling at the red rock formations. One of the big fascinations in Sedona is vortex seeking. It's a New Age thing. Those who have faith in this power think that in certain geographic locations in Sedona there are energy fields that allow you to become connected with the Earth. They can't explain why, but it is a scientific fact that electromagnetic fields are involved in metabolism, heart rate, body temperature, and blood pressure in humans. On my hike I passed several New Age believers sitting in circles, arms about one another, humming mantras, and feeling the vibes.

Oak Creek trail was supposed to be one of the locations, and sure enough, after an hour and twenty minutes of walking through sand and climbing rock formations, I found my blood pressure elevated, my temperature rising, and my heart beating at a faster rate. What mysterious stuff!

In addition to the magical vortex effects, Sedona's red rock beauty is breathtaking. The formations are incredible. Throughout my days of camping and hiking, I kept asking local people where the best place was to take pictures at sunset. They would answer in hushed tones, like it was Sedona's biggest secret, "Airport Road."

Late in the afternoon, I arrived at Airport Road to find a hundred folks gathered with their Kodak disposable cameras to take pictures. Some secret! Most posed in groups with their backs to the setting sun for their pictures. I debated telling them all they would see after developing the pictures were black silhouettes of their faces, but that's the fun of vacation photos, right?

One of the tricks I learned to taking good pictures is to take a hundred photos and throw ninety away. Another trick is to carry a tripod. It improves the photography and makes you look rather professional. There might be thirty or so people elbowing each other on a rock precipice to take pictures when I arrive and start setting up my tripod. "Oh, stand aside," they say. "This guy must be from *National Geographic*." I'm given a wide space.

The high point of my visit to Sedona was the "Day in the West" jeep tour. The tour was advertised to drive past the well-known rock formations of Snoopy, Bell Rock, and Tea Pot, among others, and also included a thrilling rock climb. How much of a thrill could that be? I wondered, as I watched Gramps and Grandma and Baby Snooks crawl into other jeeps.

My jeep was configured to carry four people in back on bench seats. Riding with me were an elegant couple from Atlanta I'll call Miss Scarlett and Mr. Butler, and another ordinary-looking couple from Ohio, Ma and Pa Kettle. I rode in the front passenger seat. The driver introduced himself as Pecos, an authentic cowboy! He was from Texas, carried a six-gun and Bowie knife, and wore a beat-up cowboy hat. Most notable were his hands, the size of catcher's mitts that looked like they had seen a lot of hard labor. Pecos was good and knew his tour-guiding patter. Talking in a slow Texas drawl from the get-go, he didn't pause for a breath the entire two-hour-and-fifteen-minute ride.

We completed the notable rock formation drive-bys and then started the thrilling rock climb. I didn't expect much but was wrong. I swear we climbed and descended some near-vertical slopes. Miss Scarlett turned white, absolutely terrified. Ma Kettle screamed bloody murder. Mr. Butler kept saying, "My Lord, my Lord, my Lord!" The four in back spent more time in the air than they did in their seats.

Pecos was loving it. I think he hit those slopes faster than normal just to hear the screams. At one point he said to me in a whisper, "Thank God for Depends, huh?"

***

How lucky I have been not only to survive my accidents, but to experience such wonders in my travels. And isn't it amazing how it hardly ever rains as I continue to stumble upon my once-in-a-lifetime memories?

Sedona sunset.

Chapter Fourteen

Are We Having Fun Yet?

When I first checked into my squadron, Patrol Squadron Twenty-One, in 1964, they had just returned from a six-month deployment in Sicily and therefore would be home in Maine for eighteen months before leaving again. Since the squadron was settling back into home-base routines and reorganizing its departments and procedures, the timing of my arrival was not ideal for becoming integrated into a new outfit.

The simple solution to what to do with me was to send me away for specialized training. Since our primary mission was antisubmarine warfare, I received orders to go aboard a submarine for ten days out of New London, Connecticut. I was apprehensive about the assignment. My eighteen-month career in the Navy had been as an aviator, and the notion of going to sea was exciting but a bit disquieting. It was not the idea of voyaging under the ocean, but I was unfamiliar with shipboard rituals and customs.

I arrived in New London at 11:00 p.m. on a rainy, gloomy night. Wandering along the waterfront past all those low black shapes, I searched for the USS *Cobbler*, a diesel-powered submarine. I found a watch officer, who directed me to the boat. After a confused conversation, it became apparent that those around at that hour did not expect me, and I was told to find a bunk somewhere; the captain would deal with me in the morning. It was not a warm welcome.

The next day I found that the captain had been aware of my assignment, and he explained the USS *Cobbler* would sail from New London to Halifax, Nova Scotia. En route we would conduct a training exercise with a VP aircraft out of Brunswick, but he didn't know from which of the five Brunswick squadrons it would be. The captain and the other officers were polite, but rather distant. I felt as if I were in a different branch of the service. It may have been the rumored arrogance of naval aviators that put my shipmates on guard, but that was not the case with the enlisted crew, who were open and friendly. I'd never seen such courteous sailors, although that might have been a prerequisite for submarine duty, where quarters were tight. Curious about my flight training, the crew wanted to know all about airplanes.

USS Cobbler

The USS *Cobbler* was a celebrity. Walter Cronkite had done a television special aboard the *Cobbler* for the show *20th Century*. A film of the program could be viewed in the ship's wardroom.

I had no problem dealing with what people assume must be claustrophobic conditions. I banged my head a few times, but swinging through the low, narrow openings between watertight compartments soon became easy. Operating the head was a challenge, since it involved turning a number of different valves, but I learned. Formalities in the submarine were relaxed; both crew and officers wore civilian shoes, sweaters, and T-shirts. There was no sense of night or day. People were either coming off or going on watch during the twenty-four-hour day, sleeping in four-hour stretches. Meals were available at all hours, although one meal consisted primarily of breakfast items.

As a conventional diesel-powered boat, we needed to spend several hours a day on the surface recharging the batteries. Whenever we surfaced, I headed for the conning tower, not because I needed to escape the confines below, but because I just enjoyed being topside. Standing in the open with binoculars around my neck to check out the distant ships, I loved the feel of wind and salt spray in my face, and gazing out at shoreless horizons. Even in rain showers, I would don rain gear and ride in the tower.

Halfway to Halifax we were attacked by a Neptune airplane. That started the exercise with the patrol squadron. To my delight, the aircraft was from my squadron in Brunswick. The submariners all think that fly-boys are show-offs and talk in a very undisciplined manner. I had to chuckle at the sneers on the bridge as the flight crew discussed the exercise they wanted to set up with the sub over the radio. "Roger, Roger, Sea Fish Alpha. Why don't you boys pull the plug, drop down one hundred feet, and hold course three-six-zero. Pop up again when hearing our signal, five Papa Delta Charlies. Over." The Papa-Delta-Charlies referred to PDCs — practice depth charges — small, harmless explosive devices.

After several aircraft-directed exercises, during which the sub held a constant underwater course for the aircraft to practice tracking, the sub was then allowed to evade using its own maneuvers and deceptions. The results were debatable, since no one was actually attacked and sunk. In their minds, I think, the aircrew decided they had won, but the sub crew probably thought, "No way. We gave you the slip."

Once in Nova Scotia, one of our flight crews from Brunswick picked me up, having flown to Halifax on a training flight. My time on the *Cobbler* had been enjoyable. I think I even persuaded one or two of the officers that not all Navy Airedales are pompous show-offs. If aviation had not been my first choice, I could have been happy as a submariner.

After my return to Brunswick, almost immediately my next get-out-of-our-hair orders were to accompany an aircrew back to Halifax, Nova Scotia. We were to be the guest of the Canadian Air Force, which had responsibility for the same mission, antisubmarine warfare, as the United States Navy. We were acting as U.S. ambassadors on a goodwill mission to reinforce relations between our two nations.

The Canadians were gracious hosts. The evening we arrived, we discovered they enjoyed their happy hour even more than their American counterparts. Bonds quickly formed. Flight Lieutenant Jerome McFadden, called Mack, was the plane commander hosting our crew. Flight officer Drew Hubbard was his copilot, and the flight engineer was a forty-year-old warrant officer by the name of Spencer Pearson. The morning after our arrival, the Brunswick crew flew as observers on the Canadian CL-28 Argus, the Canadian Air Force airplane used for maritime duties. The Canadians were extremely proud of the Argus. Since it was first commissioned in 1953, the Canadians had flown thousands of hours in the Argus and never lost one aircraft due to an accident.

The CL-28 was larger than the Neptune by forty feet in both length and wingspan. Its four propeller engines had more horsepower than the two on the Neptune, but our jet engines more than made up the difference. Their maximum airspeed was 287 miles per hour; ours was 403. Our climb rate was twice theirs, although our service ceiling and theirs were the same. The Argus had a crew of fourteen, two more than the Neptune, but the space aboard the Argus allowed both aircrews to move about with ease.

Canadian Air Force, CL-28, Argus

Taking off from Halifax, the Argus headed inland above the hilly terrain of the Nova Scotia mainland. Mack invited Lieutenant Bill Corrigan, copilot of the VP-21 crew, to come forward and take the copilot's seat in the large cockpit. I stood behind the two pilot's seats. Behind me on the starboard side of the flight deck was the flight engineer's position. Unlike our airplane, the flight engineer managed the engine controls at the command of the plane commander. We were flying at 6,000 feet, approximately 4,500 feet above the hilltops below.

Mack called back to Spencer and told him to feather the number four engine. How confident, I thought. We never performed actual engine-out maneuvers at less than six thousand feet above the terrain. The retrimmed Argus adjusted beautifully to flying on only three engines, never losing a knot of airspeed. Mack then instructed Spencer to feather number three engine. Now I was impressed. The big aircraft flew with only two engines operating on one side and lost not a foot of altitude and no more than four or five knots of speed.

When Mac called for shutting down the number two engine, I thought, Now, wait a minute, that can't be done. When Spencer brought the third engine to parade rest, Corrigan looked back at me from his right seat, as if to say, "How about that for an impressive demonstration?"

But then performance started to deteriorate quickly. Losing altitude at three hundred feet per minute, speed dropped fifteen knots as the nose of the aircraft was raised in an attempt to slow the descent.

"Power up on number one, Spence," was Mac's next command. "And bring back number four." Spencer, out of sight of the earth rising below us, was unruffled and matter-of-factly mentioned he was checking engine temperatures. The aircraft had now lost another ten knots, and Mack had traded away a thousand feet of altitude in his effort to prevent the loss of more airspeed.

The urgency in his voice could not be hidden when he directed Spencer to unfeather engines two and three and add power to four — now! Spencer, still unaware of the rapidly deteriorating situation, calmly responded to the captain saying, "Sir, number four is not up to temperature, hold on a bit."

The "backside of the power curve" is an aeronautical expression that describes our situation exactly. It refers to a graphic chart plotting power, airspeed, and lift. The curve on the chart shows that increased power results in greater airspeed and more lift. However, at the extreme low end of the chart, the curve reverses, showing how at slow airspeed, more power causes your aircraft to pitch up, losing more airspeed and lift. And that's where we found ourselves. The only way out of the situation was not to add power, but to reduce it, lowering the nose to gain airspeed. Unfortunately, we were now less than a hundred feet above the treetops, and lowering the nose would put us into the terrain.

We were going to crash.

The undeniable reality of that outcome suddenly slowed the passage of time to a surreal image. I saw everything in slow motion. I saw Mac lean forward in his seat, straining against his harness to see beyond the hilltops. Bill Corrigan, riveted to the sight of the approaching rock and wood of Mother Earth, was groping almost casually to find the shoulder harness he had neglected to fasten. I told myself that when the aircraft hit, the tail would break free, and that's where I ought to be. Incredibly, I could picture the wreckage burning yet saw the tail section torn loose and whole. As I turned to run to the rear of the plane, my last glimpse out the forward windscreen was of the tree limbs about to strike the leading edge of our wings.

The crew in the rear of the plane was unaware of the pending disaster. As I ran by, I shouted, "Going down! Going down! Hold on!" At the aft end of the fuselage, I dove for the deck, covered my head, and waited. And waited.

The sound of outside rushing air increased. The aircraft's attitude leveled. And then I heard all four throaty engines generating take-off power.

Safely back at the Halifax aerodrome, I asked Bill Corrigan what had happened. He couldn't believe it. "Those tree limbs were within inches of our wings," he said. "Just when I thought we were about to stall out, the aircraft crested a hill, and the valley dropped away below. Mac pushed the nose over, and Spencer slammed all four engines into full power, and we were able to recover."

The incident was a great embarrassment to the Canadians. We departed that same afternoon after a somber farewell. No report of the situation was ever made by our crew. However, Mac's immediate disappearance after we landed suggested he'd found himself in deep trouble.

Whew, close one.

#

Chapter Fifteen

Humor in Uniform

Formality in the military varies from service to service and from outfit to outfit. I reported earlier how submarine crews wore civilian sweaters and shoes, but that wouldn't be allowed on an aircraft carrier with an admiral aboard. Aviators are regarded by ship-going Navy men as rather cavalier and casual in their speech and manner. That may be true. The difference between Airedales and Swabbies is further differentiated because aviators had slightly different uniforms. Pilots had brown shoes they wore with their khakis, where the seagoing types wore black shoes. "Brown shoe sailors" therefore distinguished the aviation branch from the "Black shoe Navy."

Casual though we fly-boys may have been, the French Air Force astonished us with its laissez-faire attitude. In 1967 a detachment of three Navy aircrews received orders to depart our deployed base in Rota, Spain, and fly to Nimes, France, to become part of a joint antisubmarine-warfare exercise. French pilots had their basic training field at the Nimes aerodrome. We were astounded to see the French using old hand-propped spruce-and-fabric Jennys as their basic trainers. Landing at that field was like stepping back into 1920 aviation history.

French Air Force trainer, 1920s Jenny

Most of the French officers spoke English and demonstrated exceptional friendliness, but casually so. Our senior officer was a lieutenant commander, not considered lofty in rank but deserving of some courtesies. We found it surprising their most junior officer seemed comfortable calling him by his first name.

The operation we participated in included French and American destroyers, two French submarines, and French and American aircraft. The French ran the show. The submarines were challenged to transit the Mediterranean from one geographical longitude to another, using all means at their disposal to evade detection by ships and aircraft. A U.S.-directed operation of this sort would be highly organized in its disciplined operation. The French, however, treated the exercise as if they'd thought it up that morning and were happy to play it by ear.

Our aircrews just looked at each other with disbelief and shook our heads. Safety is always first priority in these peacetime operations, and flight-safety radio reports were of utmost importance. Should a ship or an aircraft experience a problem, or not report in, then the exercise is put on hold, and search or assistance becomes a priority.

Our crew's responsibility was to search a far section of the exercise area. After arriving on station, we attempted to send our first flight-safety report back to the French headquarters. No reply. The radio operator continued his attempts for the next hour, even relaying some high-frequency messages through distant listening stations. Still no reply. The time grew near when a U.S.-managed operation would start launching a search and rescue mission. At last the French operational commanders came on the air, saying rather casually, "Yes, yes, we hear you, Airmail Twelve. Do not worry, the war is not over."

What the hell was that all about?

We continued the exercise but ignored further communication attempts with headquarters. Instead, we stayed in touch with our other squadron aircraft, which had experienced the same lack of response to their calls. The squadron's new wise-ass reply to any request became, "Do not worry, the war is not over."

Incidentally, those two French submarine crews who were supposed to transit the area decided they didn't want to play and went ashore at Monaco. *C'est la vie.*

***

I spent my years in the service during the Cold War era with the USSR and USA keeping a close eye on each other. The Soviets had electronic intelligence-collecting ships, called ELINT Trawlers, stationed offshore at every Navy base from which I ever flew. They had ships just offshore from Maine to Florida, up and down the West Coast, as well as at U.S. bases in other parts of the world. The trawlers rarely were aggressive, just floated three miles offshore, monitoring Navy communications. We would fly by following every patrol, taking pictures of any new antennas they had installed. We would wave, they would wave, and that was the extent of the confrontation. Russian submarine tenders refueled the trawlers, which remained on station for months at a time.

Russian ELINT trawler

The trawler offshore at Rota, Spain, in 1965 became a different breed of cat, however. Its crew was downright confrontational. They got into the middle of our training exercises with surface ships, trying to recover practice weaponry, and would jam radio waves when we communicated with shore stations. What were we to do with these quarrelsome communists?

Our six-month deployment at the Rota base was over the month of December, so we decided to give our Bolshevik friends a Christmas gift. Really, something nice. We filled a five-foot-long sonobuoy packing tube with popcorn as a cushioning material and into it placed a fine brand of U.S. vodka, plus a variety of American beers. In order to share our American literature, copies of *Playboy* and *Penthouse* magazines were included. A sailor in the squadron found a model of their trawler sold at the Post Exchange and glued it together, painting the model in their colors, and inscribing their ship's name on the stern. The pièce de résistance of our gift, however, was a condom one of our squadron sailors stretched to gigantic proportions and wrote on the side, "Made in U.S.A. – Small."

Two aircraft made the delivery, one to drop the parachute-retarded gift, and the other to take photographs. The trawler, obviously aware something strange was going on, just circled the gift when it hit the water. Seeing the ship reluctant to pick up the package, the two aircraft departed.

The next day, when the patrolling aircraft flew by to take pictures, the entire Soviet crew was on the deck, waving the vodka bottles and holding up the magazine centerfolds. They say President Reagan gets credit for tearing down the Berlin Wall, but I think our squadron ought to be acknowledged for at least loosening a brick or two.

In 2010 I became curious about the forty-six cadets with whom I had gone through Navy preflight training in Pensacola, Florida, forty-eight years earlier. Preflight, a sixteen-week period of Officer Candidate School, included instruction in physical fitness, academics, and military training. Hardcore, in-your-face, combat-veteran Marine drill instructors conducted most of the drilling. If you haven't had the experience of a Marine DI two inches in front of your face, giving you some friendly advice, you haven't lived.

Pre-Flight Class 28-62, author top row, 7th from left

Through Internet searches and university records, I was able to locate thirty-six of my forty-six classmates. I learned six had died, some heroically in the cockpit of airplanes. The others, reached by telephone, all said, "Let's have a reunion."

We held the reunion in Pensacola. During our three nights together, each classmate was asked to give a one-minute talk on a preflight memory, a two-minute talk on his Navy career, and a two-minute talk on his career after the Navy. Everyone spoke for over thirty minutes, and it was great!

When Archie Lawson arrived at preflight, he was a five-foot, 112-pound kid, who looked like he was twelve years old. Archie came from Tennessee, and the first time he saw a flush toilet, he admitted, was when he went away to college. He made the mistake of wearing a chartreuse-green T-shirt that day, and the DI immediately christened him Green Shirt. "Green Shirt," he would announce, "I'm going to melt you down to a puddle of piss." The Navy's minimum weight requirement of 115 pounds meant that Archie had to weigh in at the base hospital every week until he gained those extra three pounds.

Archie Lawson, future Delta Captain

Archie never lost his preteen appearance; he looked then and still does today like a young Barney Fife of *Mayberry RFD*. After getting his wings, Archie flew the huge C-130, Hercules aircraft for the Military Airlift Command. At the time transporting GIs to and from Vietnam was a primary part of its mission. Archie told the story at the reunion that on the day he was evaluated for promotion to plane commander, he had an actual flight bringing war-weary Marines back from Vietnam. At the completion of the flight, the check-pilot told Archie, "Well, you did all right. Did just fine. But don't ever let those battle-bruised Marines in the back see who is flying this airplane."

Archie Lawson had just retired after forty years as a Delta Airlines pilot.

All classmates in my preflight class were Naval Aviation Officer candidates and college graduates, with one exception. We had one Marine cadet by the name of Chuck Geiger, who had just two years of college. Chuck would not be commissioned after the four months of preflight like the rest of us, but he would have to first finish flight training eighteen months later, before putting on the bars of a second lieutenant. Chuck had an impressive background. He was the son of a Marine general, but even more impressive to all us greenhorn college boys, Chuck had spent two years as a grunt in the Marine Corps. When it came to marching, saluting, and manual of arms, Chuck was one squared-away Marine.

COL CHARLES R. GEIGER, *USMC*

Geiger had one problem, however, that of staying awake in class. Days started early in preflight, with reveille at 5:30 a.m. and a full day to follow. There was an understanding in all our classes that if you became sleepy, it was permissible to get out of your chair and go stand at the back of the classroom. Chuck did that in every class.

In the military history class when Chuck moved to the back of the room, he stood in front of a door and discovered that by hooking the back of his belt over the door's knob, he could support himself even more comfortably.

About midway through the lecture, we heard a wild scrambling and beating on the wall behind us. Chuck had fallen asleep, but when his knees buckled, the doorknob kept him suspended, swinging him into a horizontal position. He flailed his arms and legs, trying to regain his footing.

The lieutenant commander teaching the class handled the incident beautifully. We learned later that he quizzed MARCAD Geiger and expressed real concern about his sleep problems.

"Is your rack not comfortable enough for you, Cadet?"

"Oh, no sir. My rack is very comfortable."

"Well, I don't believe it is, since you're dozing off in my class. I'd like you to bring that bunk up here to my fourth-floor office so I can inspect it."

Geiger recruited one of his other Marine buddies from another battalion to help him carry his bed out of the barracks, across the air station, and up the stairs to the instructor's fourth-floor office.

The LCDR was reported to say, "Looks okay. Carry on."

Try as I might, I was not able to reach Chuck Geiger prior to our reunion. I did reach him afterward and discovered that Marine Colonel Chuck Geiger had an impressive and highly decorated war record in Vietnam. When talking to him by phone, I brought him up to date on his preflight classmates. My final report was that he would be very proud of his classmates who had crashed six Navy airplanes. Chuck replied, "Well, make that seven, because I left one in Da Nang Harbor with battle damage."

***

Our squadron's second deployment to Rota, Spain, in 1967, occurred during a particularly stressful period of world affairs. Israeli/Arab friction erupted in what became known as the Six-Day War, USSR flights over American bases in Iceland and Alaska were becoming more frequent, and Soviet submarine incursions into U.S. waters had antisubmarine warfare forces on high alert. After three months of the six-month deployment, my crew had not had one day off.

Being less privileged was not totally unexpected. Of the fourteen crews in the squadron, ten of them had lieutenant-commander or full commander patrol plane commanders (PPCs). I was proud to be one of the few lieutenants to have his own crew but realized my junior rank meant I would end up with more than my share of the least-desirable assignments. Who do you suppose got the ready-duty crew assignment every Christmas Eve and on other holidays?

In Rota, after weeks with no break, I finally arranged a five-day leave for our crew. We were ready to depart for a week in Seville, when orders were received to send a crew to Sigonella, Sicily, as observers in a NATO exercise. A squadron deployed in Sigonella needed witnesses — pilots, ordnance men, radiomen, and flight engineers — aboard their aircraft to verify compliance with the exercise guidelines. It was a crummy assignment. So, guess who had their leave canceled and were sent to Sicily?

The morning we arrived in Sigonella, I told the crew we had sixteen hours before our scheduled rides as observers the next day. "If," I asked, "you don't mind missing a night's sleep, I'll arrange for transportation, and we can all go to the city of Catania for a little R and R." All enthusiastically agreed.

I called the base's Special Services and learned it only had a forty-four-passenger tour bus available for the twelve of us. "Bring it on," I said. While waiting for the bus, I made a dash to the PX, where I purchased a case of beer and bottles of wine for the sixty-minute trip.

The ride to Catania was memorable. Whenever the crew spotted a pedestrian along the road, they insisted we stop and share our commodious transportation and refreshments with the locals. Young ladies were particularly welcomed, and soon we had a full load of joyfully singing companions on the trip to the big city.

Catania was fascinating. It was founded by the Greeks, before the Romans conquered Sicily in 263 AD. It is located on the side of Mount Etna, Europe's highest active volcano. The city was full of young backpacking Europeans attracted to the city because of its sights and the opportunity to hike up Mount Etna.

I have a lasting picture in my mind of the crew, at the end of the sixteen-hour liberty, locked arm-in-arm with other revelers, strolling down the streets of Catania.

Combat Air Crew 12, Plane Commander J.F.Stark, 1st row, end right

# # #

Chapter Sixteen

Heavy Mileage

Today, with everyone and his brother running triathlons and Iron Man events, a simple marathon wouldn't get much attention, but my notoriety in my hometown following the Boston Marathon in 1979 amazed me. When my family returned home — before leaving Brian in New Jersey became known — a sign had been plastered on our garage, saying, "Welcome Home our own Jim 'Bill Rogers' Stark. Congratulations!" Notes and letters from friends and business associates were most flattering. One of the more eloquent came from the president of the local United Way, who said in part, "Your excellent participation in the Boston Marathon is a strong and stunning achievement. It gives one courage to continue believing that the individual, in our teeming super-tribal society, can still merit recognition for personal effort. Hearty Congratulations, Jim."

In a follow-up article in the *Republic* newspaper, J. D. Lewis, sports editor, wrote, "The vice–president of Kirby Risk Supply Co. said, from the first foot to the end of the race ... was screaming, shouts of encouragement and cheering. Kids with their hands outstretched trying to touch the runners ... just a sea of people. Stark said the enthusiasm of the crowd swept you along. It was like a double overtime basketball game between Purdue and Indiana. It took your breath away."

I limped around for a week after the race, feeling very heroic. Now my youngest son, Brian, runs forty and fifty miles a day across states, for weeks at a time.

So, for forty years I was a compulsive runner. I worked at it, running forty to sixty miles each week. During ten peak years I ran over 2,500 miles each year. In 1983 I exceeded 2,800 miles. That's a fifty-four-mile average for each and every week. A friend commented, "I didn't put that much mileage on my car!"

As I reflect on these schedules and routines, now in my advancing seventies, after having a hip replacement and facing a future knee replacement, knowing my running days are over, it makes me acknowledge that my obsession was extraordinary. Did I really do that?

I did. I am apparently ingrained with that kind of mania. The words "discipline" and "dedication" sound rather heroic, but I believe such perseverance is an individual orientation. I was running ten to twelve hours a week. Many people spend that amount of time getting advanced degrees. It's just a matter of commitment.

I joined the Navy after college and loved it. Why? Because it had rules and disciplines. Rules you must follow, especially for a pilot. When the light on the instrument panel blinks red, you must pull the ejection handle and bail out. That's the rule. Smart guys might want to figure out the problem and discuss the issue, or form a committee, but two seconds later they would be dead. Washouts in the Navy's flight program were not always the inept; they included a fair number of smart guys who challenged the rules. I liked rules, those I set for myself and those set by others.

One of my personal record goals was to run a 10K race (6.2 miles) in under forty minutes; that's a 6.27-minute-per-mile pace. It became my quest, but I hoped not an impossible dream. I trained running ninety-second quarter miles on a track — a six-minute mile pace — but in race after race, uphill or down, cold weather or hot, although I came within seconds of that time: 40:20, 40:06, 40:15, I never achieved that sought-after sub-four-oh! That carrot on the stick kept me training harder and longer than I might have otherwise.

Because of the detailed records I kept, the thought struck me it would be noteworthy to recognize the point at which the miles I ran equaled the circumference of the Earth, 24,859.82 miles. One night as I lay in bed, I added the mileage in my head and realized I'd passed that mark several days earlier. Oh, well, how far to the moon?

In 1980 I planned to run the St. Louis Olympiad Memorial Marathon. I hoped to set a new personal record for the marathon. My eleven-year-old son, Chris, an occasional runner, decided he wanted to run the 10K held at the same time as the marathon. I told Chris that a six-mile race would take training; if he were really serious, he needed to follow a three-week program of about three miles a day to get ready.

Chris was indeed serious and put in the mileage. When we arrived in St. Louis, a snowstorm the night before covered the streets in a blanket of white. Plows cleared the course, but I realized that with the remaining slippery areas and the St. Louis hills, I would not be able to achieve the personal record I sought. I decided instead that I would wait and try for the record at the Toledo Marathon two weeks later, but in St. Louis I would run the 10K with Chris.

When I announced my decision to my son, I couldn't understand his less than enthusiastic response. His mother deciphered his reaction. Chris had accepted this challenge as a solo experience. This was something he wanted to do himself. And now ol' dad was stepping in to make it easier for him. "Oops. Okay, Chris, you're on your own. Good luck." Nevertheless, I planned to run along, keeping Chris in sight from a distance.

I suspect that Chris knew my surging ahead in that race, then falling back was only to keep him in sight, but he did great. However, I could tell in the last couple of miles he was feeling the raw edge of exhaustion known to all competitive runners. I ran ahead to be at the finish line to watch him cross. His stride was smooth, his pace was even, and his determination was undeniable. The glow on his face reflected more than the protective smear of Vaseline applied against the cold, reflected more than the inner joy of all runners who gave it their all and did well. It reflected the pride and admiration of a father a few feet away experiencing a unique runner's high.

My oldest son, Eric, became a runner as I did, later in life as a forty-year-old. Eric runs daily and has competed in several 10Ks, mini-marathons, and, at this writing, at least one full marathon. Chris, my middle son, was captain of his college cross-country team and has run several marathons. But it's my youngest son, Brian, who puts us all to shame. I could fill this chapter with Brian's running exploits, but let me just report that he ran across the United States in 1998, unsupported, with nothing but a fanny pack with his supplies. Every day and every night was a challenge for him to find something to eat and somewhere to sleep. He published a book about his exploits while running from Cape Henlopen, Delaware, to Point Reyes, California, titled *Getting to the Point in a Dozen Pairs of Shoes*, which gives all the details.

Brian now has a goal to run across every state in the U.S.: fifty states by age fifty. As of 2015 he had crossed thirty-six of them. Brian doesn't just run the shortest distance across the states; he runs trails, off roads whenever possible. To use Indiana as an example, it's only 164 miles to cross Indiana on Highway 50, but Brian ran on the Knobstone Trail and others, running twice that distance to cross the Hoosier state.

Each summer Brian selects another one or two states to cross. I have occasionally rendezvoused with him while on my motorcycling trips. Brian documents his runs with video and makes them into professional documentaries. The last one, running across Wisconsin and Michigan, was shown in a Columbus movie theater. In the film you see me on my motorcycle, meeting Brian and picking him up at the end of a day's run, taking him to a motel. En route, much to my surprise, a camera is extended on a stick from the back seat, filming us on the motorcycle. Amazing!

Running has been a significant part of my life: proud achievements, good friends, memorable moments, and a healthy lifestyle. But I am most pleased about the mutual interest it provided for my sons and me. We've run together, I've gone on their high school cross-country camps, cheered for them at track meets in high school and in college, and marvel at their continuing running achievements. It's been a bonding common ground for us all.

###

Chapter Seventeen

Westward Ho

The wind blew fifty miles per hour, sand pelted my face like buckshot. Fortunately, my motorcycle helmet's visor covered my eyes, with its lower edge snug against my cheeks. The bike was heeled over ten degrees, leaning against the crosswind. It was hairy, but I was able to maintain my position in the highway's proper lane until those blasted trucks passed and stirred the winds into a whipping maelstrom. The blast pushed and pulled the bike, challenging the stability of its seventy-mile-an-hour speed. Slower speeds only decreased the stability. If the road would just turn ninety degrees, a direct head or tailwind would be much easier.

  I was nearing the end of the second week of my much-anticipated three-week motorcycle/camping excursion. This was the first of six extended trips I would take over the next dozen years. This venture took me west, spending several days in both Colorado and Arizona.

  The windstorm caught me right after I left Grand Canyon National Park in Arizona and headed toward a geographical area known as the Four Corners, where the states of Arizona, Utah, New Mexico, and Colorado come together. My destination for this leg a couple days later, was Boulder, Colorado, where I would spend two days with my former college roommate, whom I had not seen in forty years. But first I needed to deal with the gale-force desert winds.

  A convenience store loomed up beside the windblown highway, and I pulled in, hoping a cup of coffee might calm my nerves and perhaps give the conditions time to subside. I parked the thousand-pound motorcycle with its hundred-pound load of tent, sleeping bag, and camping gear on its side-stand and went in to decompress.

As I sipped my java, I struck up a conversation with one of the customers. "Think these winds will blow themselves out anytime soon?" I asked.

"What winds?" came the reply. "Shoot, these breezes blow like this for weeks at a time."

"Damn. Sorry I asked."

With that bit of non-encouragement, I finished my coffee and ventured out into the dust devils swirling around the parking lot. I had parked the motorcycle leaning away from the wind's direction. After restarting the engine, I struggled to raise the bike to an upright position. I wasn't able to stand it against the force of the gusts. By heaving the bike upright, I might have gotten it up, but there was a chance I would heave too much and throw the bike over on the opposite side. To lift a thousand-pound laid-down motorcycle back on its wheels would not be easy.

I swallowed my pride and returned to the restaurant to ask two patrons if they would help me get the bike upright. Once it was vertical, I was sure, I could zoom off under control. The customers were happy to help. As I rode off, I saw the two shaking their heads and probably saying, "Hope that fella don't live too far from here." I made it to Durango, Colorado, once again safe and unscathed. My incredible good luck continues to shine upon me.

What an enjoyable adventure it has been. I left Indiana with sights set on Tucson, the home of my youngest son, Brian. With two thousand miles to cover, I planned to travel 350 miles each day, camping each night with the gear I carried strapped to the Goldwing. I also wanted to stop at least once each day to visit a tourist attraction, stretch my legs, and discover something interesting.

My first stretch-break, in Missouri, was the Jesse James Museum in Stanton. Jesse eluded a posse in Stanton by hiding in the Meramec Caves, with which he was familiar. History claims Bob Ford shot Jesse, whose body then was identified by his mother and buried in her front yard. The museum says not so! Ford shot John Bigelow, it claims, another member of the gang; and Jesse actually lived to the ripe old age of 102, not dying until 1951. Believe it or not.

When I rode into Oklahoma that first night, I discovered threatening rain, so rather than risk having to deal with wet camping gear, I opted for a motel. It was June 6, the fifty-ninth anniversary of D-Day, 1944. In Miami, Oklahoma, I found the town full of camouflaged armored vehicles, colorful battle flags, and three thousand combatants adorned in military fatigues toting lethal-looking rifles and side arms. Miami celebrates the Normandy invasion each year with a colossal paintball war. Forces include dads, moms, and uniformed children, all waging war across a two-thousand-acre battlefield. I tried to imagine the briefings taking place that morning. "Okay kids. Your mother will lead the other moms on your flank. You kids center the enemy in your crosshairs and annihilate the bastards."

Only in America.

Were these examples of other sights I would discover on my travels? I wondered. If so, I was off to a good start.

Still in Oklahoma, I stopped in Claremore the next day to visit the museum of my boyhood hero, Will Rogers. He died before I was born, but I read his biography and became mesmerized by his many talents. Rogers was the celebrity every other celebrity claimed as a friend. Lincoln would have boasted that *Will*, a frequent visitor to the White House, slept in *his* bed.

Rogers died in a plane crash in 1935. The event stunned the nation. Dr. James Whitcomb Brougher gave the eulogy at Will's funeral. He said, "There are many streams, but only here and there a great Mississippi; there are many trees, but only here and there a great Sequoia giant; there are many echoes, but only now and then an original voice; there are many musicians, but only now and then a Mendelssohn or a Mozart; there are many politicians, but only now and then a commanding statesman; there are many people, but only now and then an outstanding individual."

The Will Rogers Museum is an incredible museum. Most impressive.

In Texas I camped for the night at the Palo Duro Canyon State Park. Guidebooks claim it is the best-kept secret in Texas. The state park is nestled among the red rock cliffs of a 125-mile-long, 800-foot-deep canyon. My campsite had a ramada-sheltered picnic table surrounded by canyon walls. A fellow camper pointed out the dark purple places in those walls. "That's prehistoric," he said. "That's where they find all them dinosaur bones and ancient sea turtles."

I doubt I will stay in a campsite more beautiful or dramatic than Palo Duro Canyon State Park. But maybe I'm supposed to keep that a secret.

Palo Duro Canyon Camp, Texas, 2003

After I left the panhandle of Texas and rode into New Mexico, I marveled at the unique country I was seeing. To ride a motorcycle is a sensory experience. You feel the power between your legs, the warmth of the sun on your face, and the wind on your cheek. You hear the hum of tires on the road and the slipstream of air rushing past your helmet. You smell the surroundings; the new wood of the pallet truck that just passed, the earth of a freshly plowed field, and the dung of the pig farm. You see your shadow bouncing over the roadside gullies and guardrails, and the asphalt flashing by in a blur six inches below your boots. You are part of the world around you, not isolated from it, as you are in an automobile. Placido Domingo sang "Granada" on my tape player as I swept through the mountain curves of New Mexico.

Most of the cliff dwellers in New Mexico lived in the northern part of the state, but the Gila Cliff Dwellings National Monument is in the southern part, not far from my route toward Tucson. I stopped to ask directions at a gas station, and the attendant pointed out the turn I had to make but then added, "I sure wouldn't go there. Just a bunch of holes in the ground."

I said, "I take it you don't work for the Chamber of Commerce, right?"

He said, "Chamber of whaaat?"

The Gila Cliff Dwellers, known as the Mogollon culture, were a small group (forty or sixty Native Americans) who made their homes there. The date of habitation is estimated to be the late 1270s, over seven hundred years ago. Judging by where they placed the hearth in their dwellings and the tools they used, they lived intelligent and productive lives. Fascinating!

My son Brian met me at the end of his street when I arrived in Tucson. He and his wife, Lydia, live in a lovely condominium community that is very environmentally conscious — wastewater is used to irrigate plants, etc. Brian had planned a full two-day schedule for the two of us, starting that afternoon with a hike into the Santa Catalina Mountains. After hiking up to an elevation of eight thousand feet, we could look down on the city of Tucson beyond the stately saguaro cactus. Spectacular!

The next day we headed to the Arizona–Sonora Desert Museum. Rather than a museum, it is first a botanical garden; even more, it is a world-renowned zoo and history center. William Carr created the museum in 1944 to draw attention to Arizona's unique beauty and instill pride in its natural wonders. If you find yourself in Tucson with a day to do one thing, do this museum. Beautiful. The zoo area shows animals in their natural habitat. We saw coyotes, javelinas (wild pigs), birds, lizards, bobcats, ocelots, beavers, otters, owls, snakes, prairie dogs, deer, black bears, and mountain lions.

If you think all native desert plants are dull and drab, you ought to see the color on display in the native plant section. Awe inspiring.

After five hours of touring, we headed back to Brian's home for a swim in the condo pool and to rest up for the next day's activities.

The morning started in Tombstone, where, with the discovery of thirty-four million dollars' worth of gold, silver, and copper in 1881, the town swelled with the addition of ten thousand eager prospectors. To keep the hardworking gents entertained, seven saloons operated twenty-four hours a day. Sheriff Wyatt Earp, Miss Kitty, and Doc Holliday said, "Why not? We don't have a YMCA." Fights broke out in the bars, leading to the famous gunfight at OK Corral. Boot Hill Cemetery became the final resting place for many in Tucson. My favorite site in Tombstone was the courthouse, with its backyard gallows. There, Tombstone justice was carried out at the end of a rope.

Flooded mines ended Tombstone's glory days, and the county seat was moved to Bisbee, which enjoyed even greater mining success. Bisbee would be our next stop.

Modern-day Bisbee became what San Francisco must have looked like in the 1960s. Located on the side of a hill, many unusual homes clung to its hillsides. The residents of those homes were Haight-Ashbury lookalikes: long hair and tie-dyed clothing, all wandering about in clouds of sweet-smelling narcotic smoke.

Brian:

Tombstone is the town too tough to die, and Bisbee is the town too stoned to care.

Curious about the mining history that flourished there until the 1970s, we drove through Bisbee seeing hippies, New Age folks, punk rockers, skinheads, drifters, and genuine cowboys in their ten-gallon hats. We signed up for a tour of the Queen Mine. Putting on pants and fleece pullovers — it's forty-seven degrees in the mine — we received yellow slickers and hard hats with attached miner's lights. Tour guides are all former miners who lend authenticity to the mine-visiting experience. Railcars carried us down the shafts, illuminated only by the lights on our hard hats. There in the dim light, we observed old drills, picks, and dynamite packing cases, just as the miners left them fifty years earlier. Mules had been used to haul the mine cars and were lowered into the shafts in slings, never to return to the surface. Most mules went blind in their years of darkness.

Bisbee Mine Tour, Bisbee, AZ, 2003

Kartchner Caverns, sixty minutes away, became our next visit that afternoon. It was first discovered by two spelunkers, Randy Tufts and Gary Tenen, who were so dedicated to preserving the pristine nature of this major cavern that they kept it a secret for fourteen years while taking steps to protect the cave from abuse, vandalism, and lack of respect. It is now a protected state park. Visitors go through four sealed chambers, getting skin moisture and respiration acclimated to the cave. Visitors are cautioned: keep hands only on the handrail, and touch no surface.

Brian is one of the state-employed cave guides, but this being his day off, a young lady did the guiding in an excellent manner. I've been in a number of our nation's other caves, but Kartchner is without a doubt the most impressive. Thanks to Randy and Gary and the state of Arizona for their efforts to preserve this marvel.

I bade good-bye to Brian and Lydia after my fun-filled two days and headed north for a brief stop in Scottsdale to visit the neighborhood where my stepson, Phil Risk, was building a mega-bucks home. I would then continue on toward Grand Canyon National Park, where I would camp for a couple nights. The temperature in Scottsdale topped 106 degrees. Yowie! It was time to head for high country.

There is one major highway north out of Scottsdale, I-17, and it was crammed with barely moving, bumper-to-bumper automobiles. My Goldwing's engine is water-cooled, so I didn't expect overheating problems, but balancing a heavily overloaded motorcycle, stopping and starting at a creeping pace, is nightmarish. After thirty minutes of that struggle, my patience boiled over, and I pulled onto the shoulder of the highway and zoomed past the stalled traffic, searching for an exit. I was not able to hear the protests or witness the hand signals of those trapped in their cars, but I doubt they were cheering me on.

I approached two state patrolmen in a Subway restaurant and learned that the congested I-17 traffic was an every-Saturday situation as weekenders headed for Vegas. Unfortunately, there was no alternate route that wouldn't add hours to northern travel. They did say the traffic thinned appreciatively in another ten miles and advised me to bite the bullet and hang in there.

Fortified with cool drinks and a water-soaked bandana around my neck, I returned to the highway for an additional stop-and-go ten miles. Next stop, Grand Canyon National Park.

# # #

Chapter Eighteen

A Funny Thing Happened ... on the Water

The main entrance to Disney World in Orlando is on the shore of a small lake. Many patrons park their cars in one of the mammoth parking lots on one side of the lake and take a brightly decorated ferry boat across the lake to Fantasyland's entrance.

Disney rents small boats on the lake. There are rowboats, canoes, small powerboats, and sailboats. One of the sailboats is the Hobie Cat, a twin-hulled catamaran with fabric deck stretched between its two pontoon hulls. The boat is lightweight and great sport to sail.

My family was staying at Disney's Polynesian Village on the lake in 1982, and while the kids enjoyed the attractions in the theme park, I decided to rent a Hobie Cat and go sailing. The authoritative young man handling the boats at the dock asked me in a challenging manner, "You ever sail one of these cats before? They're trickier than regular sailboats, you know."

"Oh, sure, sailed them many times." A slight stretch of the adjective "many" perhaps. "Few" would have been closer to the truth. I was aware the Hobie could get up on one pontoon in moderate winds, but you had to be careful, because if it went over, the boats were extremely difficult to turn back right side up.

I was having a great time sailing from one side of the lake to the other. The passenger ferry would pass periodically, and I felt like one of the Disney attractions as scores of patrons lining the upper and lower decks waved and shouted enthusiastically.

I'm certain the ferry ran on a track under the water, rather than actually being steered by someone on the bridge. Typical Disney automation. "Hi, I'm Captain Willie and I'm taking you to Fantasyland. Please take younger children by the hand and have a wonderful day."

Because of the ferry's lack of maneuverability, I gave the big boat plenty of room whenever close. On about my fifth crossing of the lake, I became focused on a loose rudder connection and wasn't paying attention to where I was going, until suddenly hearing the shrill sound of the ferry's whistle. When I looked up, I realized we were on a collision course. I threw the tiller fully to one side, and the Hobie Cat reacted immediately, making a 180-degree jibe.

Unlike "coming about" into the wind, a jibe-turn brings the wind over the stern, and although it's a legitimate way to change direction, it must be done carefully, or the mainsail will swing forcefully from one side of the boat to the other. I hadn't prepared for the hasty turn and, in full view of several hundred onlookers, saw the boom swing by within inches of my head, then hit the mainsheet limit, and gracefully pull the boat bottom side up.

When I splashed into the water, I had the presence of mind to surface quickly, smiling and waving to my audience on the ferry as it steamed past. Half of them no doubt thought I was part of Disney's planned entertainment.

So now what? As mentioned, getting a Hobie Cat back on its feet would not be easy. You must somehow get the sail off the mast, then hopefully, by standing on one of the pontoons, pull a halyard connected to the top of the mast and heave the boat upright. A lot depends on the wind, your weight, and luck.

I struggled without success for twenty minutes before a rescue boat arrived from the docks. The same young man who had challenged my experience in sailing drove the boat. "Thought you knew how to sail," he said.

"I was doing fine until that damn ferry snuck up on me. Rather than criticize, why don't you grab the end of the mast and help me get this thing back right side up."

He did, and I continued my sail. Knowing my Disney dock dictator was watching, I even got the Hobie partially up on one pontoon, just to prove I knew what I was doing.

I speculated that sometime that afternoon two families would meet in the theme park, talking about the wonderful time they had that day. And one family would say, "Didn't you just love it when that guy in the sailboat tipped it over right in front of the ferry?"

***

When Michele and I married in 1992, Michele had not had much sailing experience. For that reason, I signed the two of us up for the Annapolis Sailing School on Chesapeake Bay. The five-day school trained on twenty-four-foot Rainbow sloops. Even though I would not be considered a beginner, I looked forward to the classes, knowing there was always something to be learned about the avocation I loved.

When we arrived in Annapolis, the school informed our class of twenty-four enrollees that it was school's policy that husbands and wives not be allowed to sail together. It was their experience that the dominant spouse always took the lead, and there would not be equal learning. I understood that but insisted, because of our honeymoon trip, we would not be separated. The school acquiesced. There were other husbands and wives in the group, and none of the other couples minded adhering to the rule; in fact, most understood the benefits of the edict. Classes started in the morning with ground school, followed by sessions on the water on the dozen boats the school owned for training.

Both Michele and I thoroughly enjoyed the instructors and our classmates. The rainbow boat was a stable, fixed-keel boat, with a roomy cockpit, large enough for six people to sit comfortably. On the last day of class, the rules of sailboat racing were introduced, and we had several races, which I found exciting. I discovered the key to winning an officially conducted sailboat race, with its timed starting gun, was to be in an advantageous position when the gun went off. Our boat's crew of four won most of the races.

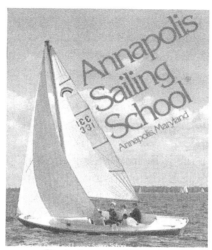

Rainbow sailboat

Chesapeake Bay is a busy body of water with a great variety of boats, including big tankers and cargo ships in addition to numerous pleasure boats and private yachts. Naval Academy sailboats were seen on the bay nearly every day. Knowing the rules of the road to avoid collisions was important. Generally speaking, ships under power are considered more maneuverable than sailboats and are supposed to yield to a boat under sail. However, the skipper of a small sailboat would be foolish to sail into the path of a large tanker, expecting it to alter its course.

When two sailboats approach each other, it is the sailboat on the port tack (wind coming over the left side) that must give way to a boat on a starboard tack. If both boats are on the same tack and converging, the boat windward (closest to the direction of the wind) must give way. During one of our races, we were sailing into the path of an eighty-foot sailing yacht. The yacht was on a starboard tack, and we were on a port tack; therefore, we were obligated to give way. At the last minute, I turned our boat across the bow of the larger boat, now putting our small boat also on a starboard tack. The large yacht, upwind of us, now had the obligation to stay clear. As it passed within feet, a crewman came to the rail, looked down on us with a sneer, and said, "Very tricky."

As graduates of the Annapolis Sailing School, Michele and I were permitted to return to Annapolis and rent one of the rainbow boats whenever the school was not in session. We did that several times over the years. Chesapeake Bay and its surrounding historic communities of St. Michaels, Deal Island, Easton, Oxford, and Tilghman Island are some of our favorite places to visit.

***

In 1994 I read an ad in a sailing magazine announcing that the Annapolis Sailing School had opened a facility on Tampa Bay and offered a weekend cruising course. Looking for a spring break adventure, we signed up. Each participant would be given one of the fleet's cruising boats, and the flotilla would sail in the fifty-mile-long Tampa Bay.

When we arrived, the person in charge told us we would be sailing a twenty-six-foot Pierson sloop. That pleased us, for a Pierson is a fine boat. "How many other boats are in the flotilla?" I asked.

Pierson sailboat, Tampa Bay, 1992

"Well," he said, "you're the only one. And since that's the case, I'll just ride along with you."

"Oops. No, you're not. We expected to be alone on a boat; otherwise, we'll just cancel."

"Okay, no problem," he said. "I can take the utility boat. I'll lead the way and show you the harbor where you're going to spend the night."

"That'll work."

It was a lovely day for sailing. Winds blew twelve to fifteen knots, with temperatures in the low seventies. The instructor motored by several times during the day and took pictures of us under way, but otherwise we rarely saw him until we arrived at St. Petersburg Marina.

The instructor met us at the dock and said, "I don't feel I have taught you anything this weekend. Would you like to learn some tricks about docking?"

"Sure, always eager to learn new sailing tricks."

The instructor came aboard and motored around the marina looking for the tightest dock space he could find, parallel docking the boat with no more than inches of clearance forward and aft of boats already tied up. We practiced his maneuvers until we could also slide into a space appearing to need a shoe horn to pull it off. Boat owners already secured close by watched in awe and some concern as we snuggled in among them.

Our flotilla of one was one fun weekend.

∗∗∗

Later that same year, Michele and I attended a business convention in San Diego. The meeting's hotel sat on the water's edge of San Diego Harbor. As we walked to the various business activities, we watched sailboats in the harbor playing catch-me-if-you-can with the seals and harbor water birds. We had to get out there.

The afternoon the meeting ended, we rented a twenty-two-foot sailboat from the marina next to our hotel and went sailing. San Diego is reported to have year-round perfect weather, and that afternoon was a good example of it.

After an hour of exploring the boundaries of the harbor, Michele suggested we take the boat out into the Pacific. I could see whitecaps on the ocean and wondered how long it would take Michele to decide a twenty-two-foot boat was no match for ocean waves, but out we went.

The surface conditions weren't as rough as I'd thought. Even more interesting were the large groundswells that lifted our boat twenty feet above the troughs between each crest. The swells were gentle and widely spaced. It was a smoothly docile, up-and-down elevator ride.

Then I spotted an aircraft carrier returning to its San Diego Navy yard. Of course, as a boat under sail, we had the right-of-way over a ship under power. Yeah, right! I did a quick come-about and headed in a direction I thought well clear of the United States Navy.

The interesting part of this encounter was how the carrier would disappear from view when we were in the trough of the swells and then reappear as we rose on a crest. The carrier was so huge we didn't seem to be increasing the distance from it. It was always just there, charging after us.

After thirty minutes of now-you-see-me, now-you-don't, I could tell the bird-farm, as we pilots called them, was not heading for our harbor but steaming toward a different destination.

I often wonder whether there had been a conversation on the bridge of the carrier about our little sailboat. "What's that idiot doing out here in the ocean? Probably one of those crazy Navy pilots."

***

Playing chicken with an aircraft carrier reminded me of another occasion when I did something similar with an ocean liner. We had been visiting a new grandchild at the time in Newport, Rhode Island. We stayed with Michele's son and wife for a few days, and I thought it would be fun to rent a sailboat and sail on Narragansett Bay. Michele was busy teaching the new mother how to care for her baby, so I went alone.

The sailboat was small, fifteen feet or under. Winds were fresh and I sailed with the mainsail only to keep the boat under control. There were many other craft on the bay that day, pleasure boats, commercial fishing vessels, and a large ocean liner that had anchored in front of the Newport waterfront. I sailed past it, marveling at its enormity when close abeam.

After a couple hours of fun, I thought I'd make one last pass by the cruise ship before calling it a day. This time I decided I would pass by its bow. The gusty winds occasionally heeled my boat over close to the tipping point, and I had to let out the mainsheet to keep from upsetting. Trying to maintain my heading just past the liner's bow, I thought it strange that I needed to keep turning to the port side if I was going to get past the bow. Then it struck me — that ship is no longer anchored; it's moving.

Okay, you win, big boat. I'm a chicken and I'm outta here, much to the relief of those on the ship's bridge, I'm sure.

***

In 1995 Michele and I had a delightful vacation in San Francisco. We did the wine country, visited Muir Woods, spent some days on Monterey Peninsula, and rode along the Big Sur. One of the other highlights involved renting a sailboat out of Sausalito on the north side of San Francisco Bay. We circled Alcatraz, sailed past Fisherman's Wharf, and decided to sail under the Golden Gate Bridge. Winds and currents are always present in the bay, one of the reasons they felt Alcatraz was escape-proof, but we remained confident we had all the elements under control.

As we approached the bridge, the wind blasted our faces and howled in our ears. The boat heeled over thirty degrees, salt spray flying over the deck. In the shadow of the bridge, we could look up and make out what appeared to be miniature toy vehicles crossing above us. The plan was to come about just the other side of the bridge. It wouldn't take long to get there at the speed we were sailing. So how come we had been in that shadow for the last fifteen minutes? We were not moving! Our seven-knot sailing speed was matched by the speed of the current pouring into San Francisco Bay. "Stand by to come about! Helms alee!"

Golden Gate, San Francisco, CA, 1995

***

The following year, we again found ourselves in San Francisco, this time attending a business convention. A fellow employee, Jim, and his wife, Leslie, also attended and wanted to stay a few days after the meeting and have us show them the sights of the area.

Jim and Leslie weren't sailors but thought the idea of sailing on the bay sounded wonderful. Chartering again out of Sausalito, the first thing they wanted to do was sail under the Golden Gate Bridge. We sailed out of the marina on the north side of the bay, turned westerly, and whap! The boat immediately heeled over forty-five degrees, facing those head-on winds. Jim and Leslie disappeared below deck and didn't reappear until an hour later when we sailed downwind behind the wind-shielding island of Alcatraz.

***

In 1996 Michele and I decided to spend Christmas in the Keys of Florida. *Sailing Magazine* had ads for several homes for rent that included a sailboat. We made reservations for one located on Conch Key, about midway between Key Largo to the north and Key West at the end of the keys. Conch Key, the very smallest on the keys, had a population of forty-four permanent residents.

Our sailing pursuits have presented many other interesting discoveries, such as when driving south that December, we stopped in Ashville to visit the famous Biltmore Estate. The mansion is huge, with over 250 rooms. Its size is not measured in square footage but in acres: four acres, actually. Of all its impressive facts and features, I found it remarkable that the owners, the Vanderbilts, had so many guests that their two hundred servants were kept busy in four laundry rooms each day, two rooms for washing, one room for drying, and one room to do the dyeing. Dyeing? What, did they change colors to match their guests' preferences?

The Biltmore was decorated in its Christmas finery that December and looked particularly festive because of the rare snow covering the ground in North Carolina. It was a fascinating visit.

Conch Key, we discovered, was a fishing village, so any thoughts of white sandy beaches would be found elsewhere. Lobster traps, old boats, and fishing shacks described most of what was present on Conch Key. Our cottage wasn't so bad: an elevated bungalow at water's edge, with a twenty-two-foot Catalina sailboat tied below.

There are many sights to enjoy and restaurants to visit on neighboring keys. Marathon was just minutes away to the south, with Islamorada Key to the north, but we were anxious to check out our sailboat. The boat was moored on the gulf side of the keys. To get out into the Atlantic, we would need to sail down to one of the bridges for access. The forecast, however, indicated waves in the Atlantic of four to eight feet, so we thought it best to stay on the gulf side.

With Michele at the helm, we spent a pleasant hour sailing downwind in gentle breezes. Passing homes and docks on shore, I was surprised at how fast we seemed to be moving. That's when I noticed the lobster trap buoys we passed had a wake streaming in front of them that made them look like they were water skiing. Whoa! Now I get it. It's low tide and the current is rushing out of the Gulf. There must be four or five knots of current carrying us along in addition to our own four or five knots of speed.

Sailboat in the Keys, 1996

Doing an immediate turn to reverse our direction, we discovered those gentle breezes were no match for the current as we continued to be carried along, only now in reverse. We had three choices if we wanted to return to our bungalow: one, drop anchor and wait for slack tide, two, wait three hours for the current to reverse, or three, start the outboard motor and hope that it could push the boat faster than five knots. We chose number three and successfully returned to our cottage on the bay.

On other sails from our bungalow, now aware of the tidal considerations, we planned our sailing accordingly. On one of those days we sailed north in slack conditions, knowing that when the current reversed later, it would carry us home. Later, busting along, we immediately realized the current had its own plans and was carrying us at an angle away from our desired heading. We had to point the boat forty-five degrees away from our course in order to crab onto the desired heading. Sliding sideways was not like sailing in Indiana.

Our Key West holiday included deep-sea fishing, wonderful sunset evenings, scrumptious seafood restaurants, and visits to Bahia Honda Beach national seafront with its interesting snorkel dives over the coral.

On the way home, on the spur of the moment, we decided to spend New Year's Eve at Disney in Orlando, and witness its hour-long New Year's fireworks display.

Christmas in Florida, not a bad way to spend the holidays.

# # #

Chapter Nineteen

Civilian Aviation

While in the service, several of the other newly designated Navy pilots and I went the extra step of taking a written test to qualify for the civilian FAA licenses of private, commercial, and instrument- rated pilots. After separation from the service, I became eligible for the GI Bill, with additional aviation ratings being offered as part of those benefits. For no other reason than getting free flight time, I elected to train as a certified flight instructor. I didn't plan to become an actual flight instructor but, as part of the course, decided to teach my oldest son, Eric, to fly. It was 1974 and he was eight years old.

Eric became a good student. In addition to learning how to take off and land, fly S-turns over a railroad track, and other primary maneuvers, he enthusiastically studied the flight manuals and other related instruction materials. Besides, it gave me a wonderful opportunity to spend time with my son. We would typically fly to a nearby airport to practice landings and have lunch while discussing schoolwork, friends, and non-flying subjects.

Unfortunately, our flights together stopped when my company transferred my family to a different part of the state to open a new branch office. Due to increased business activities and Eric's refocusing on a new school, we never got back to his flying lessons.

Years later, after my son's college graduation and after getting his doctorate in music, he decided to pursue a private pilot's license. On the introductory flight at Eagle Creek airport in Indianapolis, the instructor asked, "You ever flown before?"

"Yes. My dad gave me some lessons."

"Oh, when was that?"

"When I was eight years old."

"Really?" the instructor said with a chuckle. "So, do you want to try making the take-off?"

"Okay," said Eric. He added power, held the plane precisely on the centerline, climbed to altitude, and trimmed the airplane for hands-free level flight.

The instructor, open-mouthed, said, "You were how old?"

Eric got his license and now volunteers with Angel Air, transporting medical patients to distant hospitals for treatment.

***

One of the joys of flying is the breakout when making an IFR (Instrument Flight Rules) departure. FAA says a standard instrument-rated pilot must have at least a three-hundred-foot ceiling to take off in IFR conditions. When Eric and I started a cross-country adventure to the West Coast, those were the conditions on our departure from Columbus. It was raining. Gray clouds draped themselves over the treetops. The weather forecaster said we had the ceiling we needed, and tops of the clouds were at six thousand feet. Mom was there to see us off, but nervous. She urged us to wait for more favorable conditions. "This is no problem," I assured her. "We'll call you from Hot Springs, our first destination."

The liftoff was normal and then almost immediately we were sucked into a world of gray. Some turbulence added to the discomfort of being dependent on the gyro-horizon, turn and bank indicator, and radio instruments. However, hundreds of hours of instrument flying taught me to trust these lifesavers, regardless of the confusing signals you get from your inner ear or the seat of your pants. Ascending into the rain clouds, our instruments confirmed "on heading, on course."

We climbed through four thousand, five thousand, and then the gray started to lighten. A break in the clouds appeared, then wham! We broke through that final mist, and there we were in dazzling sunlight. Whew, time for coffee and a cigarette. I haven't smoked for forty years, but I still remember the pleasure of that first puff after those breakouts all those years ago.

***

An equally pleasurable experience is when flying an obscured instrument approach to landing. Flying toward earth with nothing but your wingtips visible takes faith. Whether it is a precise Ground Controlled Approach (GCA) or the less-than-exact VHF Omnidirectional Range approach (VOR), which gives no altitude indications, you focus on the needles, trusting them for guidance. When you break out with a ribbon of runway welcoming you home, it's a deliciously satisfying moment.

Flying home from a business meeting in Danville, Illinois, I had my branch managers from Bloomington, Bedford, and Columbus in the airplane with me. Had I known the weather was going to turn as disagreeable as it was, I wouldn't have flown. Our flight over to Illinois was fine, but on the return, low ceilings and precipitation suddenly became part of the winter night sky. Ron, my Bedford manager, was the least-comfortable passenger on that flight. He would have preferred to have a wisdom tooth extraction. I made the mistake of tuning in the air traffic controller's communications on the plane's speakers, thinking my passengers might enjoy listening to them. Big mistake. Especially when the controllers started warning us of icing conditions. Ron, in front, was perched on the edge of his seat, eyes boring through the windscreen, trying to see beyond our quarter-mile visibility.

We were finally turned over to Indianapolis Approach Control, which would handle our landing approach, since the Bloomington tower personnel had secured for the day. We were cleared for a precise instrument approach, which gave us both heading and altitude indications. I told Ron, "Just relax. A couple minutes from now, we're going to drop out of this crud, and before you you'll see a long row of runway lights marking our arrival." Ron seemed to relax. As we descended, on glide slope and on heading, I could tell the clouds were starting to thin. Passing through five hundred feet, we were about to break out, and then before us … what? A black hole! Where is that runway? We were three hundred feet above the ground and needed to climb fast!

Runway lights after hours at Bloomington are turned on either by the pilot keying in a certain frequency on his radio or, when on an instrument approach, by Indianapolis Approach Control. Indy had neglected to do that. It was their fault, and they sincerely apologized.

After executing a missed approach, we were again turned onto final and landed safely. Ron, however, had had enough and never flew with me again.

Hairy moments were not just confined to the Navy, civilian aviation had its moments as well. One such experience although somewhat perilous, included an element of humor. At the time, eight of our branch offices, one in Illinois and the others in southern Indiana, reported to me. I used my airplane to make regular visits to each of the locations.

Whenever one of my branch offices would set a monthly sales record, an event that occurred at one of the businesses several times a year, I would put on a tuxedo, order a congratulatory sheet cake, buy a case of champagne, and host a celebration where I cooked hamburgers on a grill.

I was flying to Vincennes, Indiana, for such a festivity when I flew into an intense thunderstorm. Air Traffic Control tried to vector me away from the strongest cells, but I tossed about like a tissue in a wrestling match with a leaf blower. The plane gained and lost two thousand feet of altitude regardless of my efforts to control it. That's when it struck me as funny. What a classy way to die. When they find my wreckage in some farmer's field, they'll discover a formally dressed pilot, covered with cake frosting and saturated in champagne. What a way to go!

I'm pleased to report the celebration came off as scheduled, with no one the worse for wear other than an extra-loud pop when uncorking the champagne.

***

As vice-president of my company, I had flown my Cessna 182-RG—Skylane to Chicago for a business meeting. The Skylane was a great airplane. Although only a four-seater, it had retractable landing gear, a powerful engine, and top-of-the-line navigation instruments. After my meeting I was going to leave Palwaukee Airport in Chicago and fly to Purdue University Airport in Lafayette, Indiana, where I would join friends for a Purdue football game. Palwaukee Airport, just a few miles from O'Hare, required air traffic controllers to be extremely busy coordinating the Palwaukee/O'Hare arrival and departure traffic.

C-182 RG, Cessna Skylane

I filed a flight plan, did my engine run-up, and waited at the end of the runway for my departure clearance and climb-out instructions. And I waited and waited. Suddenly, I was aware of a disquieting urgency. I needed to use the bathroom. To cancel my flight plan and return to the flight line would mean being sent to the end of the queue, and I might miss the opening kickoff. Locking my knees together, I stubbornly prayed for some kind of miraculous solution. Then came the call: "Sierra November, Seven Three Six, depart runway two-seven, climb to two thousand five hundred, maintain runway heading ... and expedite, over!"

"Roger, Departure, runway heading, two-thousand five hundred, Seven Three Six." After I reached the assigned altitude, then began the rapid-fire instructions of additional climb and heading changes. "Come to heading one niner five, climb to and maintain three thousand, over." That was followed almost immediately with another heading change.

At this point I was turning blue with pain and pressure. I had to go! Do I just do it in my pants and worry about cleaning up later? On long-distance flights, I carry a relief bottle for just such emergencies, but not today. What to do? Then it occurred to me that in my garment bag, hanging back of the rear seat, eight feet away, was a plastic laundry bag over my suitcoat. If I could somehow climb over two seats, unzip my garment bag, and get that plastic sheet, I might be able to shape it into a waterproof receptacle to contain my urgent release. My plane had an autopilot, and it was a simple thing to change headings and trim for climb with the twist of a knob, but communications were constant and must not be ignored.

How I did it, I'll never know. Thanks to a lengthy cord on my headset and lip-mike, I was able to stay connected, control the autopilot, climb over the seats, unzip the garment bag, tear loose the plastic wrap, and get the job done without spilling a drop. "Roger, Departure, Seven Three Six, coming to four thousand, over. Ahhhh."

When the lineman taxied me to the parking spot in Lafayette, my first action was to open the door and anoint the ramp with my pee-pee relief.

***

I frequently took my family on flights with me. My wife was not an enthusiastic flyer, although she went, but my three sons thought it was cool. They enjoyed going places, like Pigeon Forge in Nashville, Tennessee, or to visit grandparents in Florida, but it always surprised me how the droning of the engine immediately put them to sleep. We would be flying over beautiful scenery across America or over transparent green waters of tropical islands, and they would be totally zonked out.

The relief bottle I carried on long-distance flights, well, the boys were dying to see me use it. But they missed it every time. We flew to Mackinaw Island one August after spending a week at an air show in Oshkosh, Wisconsin. As we approached the airport, I aroused all three, saying, "Wake up guys. Check it out. No motor vehicles are allowed on this island, only horse-and-buggies. Look at all the bicycles down there. Oh, by the way, you missed me peeing in the bottle."

***

Eric and I had a wonderful cross-country flight when he finished his first year of college. Knowing this might be the last time dad and son would spend extended time together, I was elated. We left Indiana with bicycles and camping gear in the back of the plane and flew to Hot Springs, San Antonio, Mexico, the Grand Canyon, Palm Springs, Santa Catalina Island, Death Valley, Salt Lake City, Nebraska, Illinois, and home. Incidentally, there are stories to be told about several of those stops that I'll get to later.

When we landed in Death Valley, the surrounding scenery appeared just as you might expect a place with that name to look. Tumbleweeds blew across the runway in blistering hot winds. The outside temperature gauge on the aircraft read 130 degrees. No one attended the ramp, and there were just two weathered, woebegone old Cessnas tied down there. There was a shack with a telephone to call for a fuel truck, and a restroom. While I secured the airplane, Eric used the bathroom. He emerged saying, "Dad, you can't believe how hot that water is in the toilet!"

It didn't occur to me at the time to ask him, but now, years later, I often wonder, How did he know that?

#

Chapter Twenty

Chasing the Sun

It was 2011 when my youngest son, Brian, received a call from a trail promoter in Nevada. "Would you be willing to attempt to set a record by running five hundred miles across the state and do so in perhaps just ten days? It would involve crossing sixteen mountain ranges for a total climb of 34,000 feet."

"Sure," he said. "Sounds like fun."

Nevada was looking for events to publicize to promote itself as a recreation destination.

Brian is no stranger to long-distance running, having run across the United States in 1998. Since that time he has crossed more than thirty states in his quest to cross all fifty before his fiftieth birthday.

To run fifty miles a day over demanding terrain would take a support crew, and Brian had them in place. Two members were a father-and-son team, Ted and Trevor Oxborrow, who, with mountain bikes and four-wheeled vehicles, would mark the route across the wide prairies and dense forests. They would also set up rest shelters, shaded from the sun, every ten miles. Brian's friend Hewett, an EMT from Louisville, would also assist, ready to provide first-aid services if needed.

I decided this would be an excellent excuse to ride my motorcycle out West, rendezvousing with Brian at occasional highway crossings and use the days in between to visit Grand Teton and Yellowstone National Parks in Wyoming.

Dodge City, Kansas, 2011

On the ride to Nevada, looking for opportunities to break up the long hours on the road, I stopped in Kansas to tour wild and woolly Dodge City. Dodge City was originally named Buffalo City, a name with a sad history. After the Civil War, there were estimated to be sixty million buffalo roaming the western plains. They were the means of sustenance for the Native American Indians. Indians derived food from the meat, clothing from the skins, and weaponry from the bones. However — and here's the sad part — the white settlers regarded the Indians as a problem, because the white man wanted the land they occupied. Our government, in its infinite wisdom, decided the way to solve the Indian problem was to kill off the buffalo herds. The government invited hunters to the plains and paid them two dollars per hide. Not bad wages when you consider a hunter could kill 150 buffalo in a day. Old photos of mountains of buffalo hides attest to the effectiveness of the slaughter. By 1880 barely a thousand buffalo remained in all the Northwest.

Buffalo City is where the hunters came to start their hunt and celebrate their kill later with whiskey, women, and song. Boot Hill Cemetery gained many a reveling buffalo hunter over nothing more than a cross word or a threatening glance.

Once the buffalo were gone, the railroad came to town, now named Dodge City. Cattlemen drove their herds to Dodge for transportation to eastern markets. Rough-and-tumble cowboys had manners not much better than the buffalo hunters', so wild times continued.

The frontier town of Dodge City was re-created using many original parts of old buildings and artifacts in the reconstruction. It is well done and worth the price of the visit.

When I entered one of my favorite states, Colorado, I found a campground near the Royal Gorge Bridge and Park. The bridge is the world's highest suspension bridge. Built in 1929, it is renowned the world over, according to the boastful brochure I picked up. There's a tourist park built at each end of the bridge and brochures say it is a "must do."

The bridge was built simply as an attraction, similar to the Eiffel Tower. But it is truly amazing. Eight hundred feet above the Arkansas River, it's eighteen feet wide, with a suspended wooden roadway that sways somewhat under the weight of traffic. Most visitors walk across, but vehicles are allowed. It's a "butterflies in your stomach" experience to look over the railing at the slender thread of river barely visible below. Both ends of the bridge are crowded with attractions. The world's steepest inclined railway may be your choice for fun, or perhaps you'd like to take the aerial tram cable car across the gorge, or ride horses up the steep trail path. Or perhaps you would like to strap yourself in for an elastic shot with bungee cords 1,300 feet over the gorge below, or do the petting farm, or watch a cowboy's gunfight. Or just maybe enjoy one of the half dozen stage performances of magic shows, musical presentations, or comedy routines. Lots to do and see.

Royal Gorge Bridge, CO, 2011

It was still early afternoon after the bridge tour, and I went looking for other activities. Zip Line Tours of the Royal Gorge has zip lines strung across the hillsides of the surrounding area. It looked like fun. The harness you wear is attached to a small trolley that slides along the cable. Nine zip lines make up the Royal Gorge installation, some a half-mile long. The tour takes nearly three hours, since you need to walk from the low landing point to the next-higher launch platform. Uphill climbs for three hours become part of the price you pay for the fun.

The next day when departing Colorado and entering Utah, I realized I was a day ahead of schedule to meet Brian. Then I saw signs for Arches National Park in Moab. Perfect. I'd find a campground, tour the Arches that afternoon and then again the next morning, and still arrive in Baker, Nevada, well ahead of Brian's arrival. A sweet young ranger in a Yogi Bear hat highlighted what I should see that afternoon and also the next morning.

Oh my God! Talk about spectacular! Those hundred-million-year-old rock sculptures are a challenge to describe. I have never seen a sight like it, ever! The red rock towers rise hundreds of feet in the air, so delicately balanced one wonders how they remain upright in even the slightest breeze, let alone through centuries of violent thunderstorms. The colors are a majestic watercolor, changing tint and hue depending on the angle of the sun. Geologists explained the formations by telling us that eons ago this area lay atop a salt bed. Over the millions of years of floods, winds, and oceans, the debris was compressed into rock, perhaps a mile thick. The rock, much of it sandstone, began eroding, resulting in the towers, formations, and arches we see today. It's still happening; it's just that a process of a thousand years isn't much compared to 100 million years. Most tourists wouldn't notice the change from one day to the next.

Arches National Park, Utah, 2011

I rode back to the park the next morning to do the second half of my Arches visit. The skies were clear, and the morning sun made the formations glow with a golden brilliance. (I know I need to tone down the superlatives. After all, adding "very" before such words as "magnificent," "overwhelming," or "fantastic" just seem a bit trite.)

Needless to say, it was an incredible sight. If a painter tried to paint these formations from imagination, he couldn't do it. The three-hundred-foot formations wouldn't be that slender. The Balanced Rock wouldn't be larger than the base on which it sits, or be off center, as it is and has been for millions of years. And that arch wouldn't support itself in one good storm, let alone through centuries of them. All around me, people were taking pictures and shaking their heads in wonder and disbelief.

I spent the morning marveling at formations with names like the Park Avenues, the La Sal Mountains, the North and South Windows, the Balanced Rock, the Lower Delicate Arch, and the Skyline Arch. I hiked through the Devil's Garden to see the Landscape Arch, the largest arch in the park. It was very ... well there I go again. It was worth the hike.

I arrived in Baker, Nevada, checked into the motel, and sat outside waiting for Brian's arrival. First to appear were his support crew, father-and-son team Ted and Trevor. Ted, the father, is the Nevada coordinator of the American Discovery Trail, the route Brian will run across Nevada. His son Trevor has a company called Nevada Trailhead Expeditions that organizes cycling, hiking, and horseback tours across Nevada. It was Ted's idea that Brian make this run, setting a record for the on-foot crossing, and then to hold an annual competition to break the record. The state park organization joined in supporting the Oxborrows, because they feel Colorado and Arizona receive all the publicity when it comes to tourism. Ted and Trevor have been on the trail this past week marking the route with flags. The trail exists, but some of the turns and routes are unclear because of cattle trails, overgrowth, and unauthorized ATV tracks.

After Brian arrived the equipment was spread out over picnic tables to check it over. The amount of camping paraphernalia, clothing, medical supplies, and nutrients that were assembled for this effort is amazing. Included in the gear are snowshoes for the high mountain passes, and large bags of nutritional powders, pills, and juices, providing the electrolyte and energy sources Brian will need in making such a demanding effort. A sponsor is providing these products. Brian made his first run across the U.S. back in 1998. He was virtually unsupported then, with nothing more than a fanny pack. My son looks at this pile of equipment now and just laughs.

We all had a last supper together (not in the biblical sense!) and topped it off with a celebratory cake provided by the motel's restaurant. Tomorrow's plan was to gather at 4:30 a.m. and ride for five miles to the Nevada state line for Brian's start. I would accompany the crew for the sendoff, then return to the motel to pack up and head north toward Salt Lake, the Grand Tetons, and Yellowstone. I'd be in touch by cell phone when communication was possible and get back to Nevada a couple days before Brian's planned finish at Lake Tahoe.

The alarm buzzed at 4:30, and we assembled in the dark to drive Brian to his starting point. Trevor and I took pictures of Brian at the state line, standing before the American Discovery Trail banner.

Then off he went with the sun rising behind him. It was a beautiful morning. The brilliantly orange sun was now fully exposed above the eastern mountains and gave the entire roadside a golden glow. "Good luck, Brian. We're very proud of you, no matter what happens over the next ten days."

Brian replied in his typical casual manner, "Thanks, Dad. I'll just run this way for a while and see what happens."

When I rode into Jackson, Wyoming, near Grand Teton and Yellowstone National Parks, I paid a visit to the visitor's center to ask advice about camping. The ranger suggested I set up camp at Colter Bay Villages between the two national parks and stay in one location while visiting both parks. Since the parks are only fifty miles apart, that sounded like an excellent plan.

When I first caught sight of the Grand Tetons, my reaction was "Wow!" Now, there's an expression Ernest Hemingway never used. How creative! How descriptive! Well, I just can't help myself. It's just Wow! Very *very* Wow!

Chasing the Sun

Brian, start of 500 mile run across Nevada, 2011

When I told my French wife, Michele, in an e-mail that I was going to the Grand Tetons, she wrote back saying, "You know what Grand Tetons means in French, don't you? Big boobs!" Therefore, I was expecting two sharply pointed pinnacles, but actually there are a number of them.

When I arrived at Colter Bay Villages, I checked in, showing my lifetime golden age pass to the national parks. What a good deal that pass has been. I bought it ten years ago for ten dollars and never have to pay to enter a national park. My good friend, Herb Hoover, is so embarrassed by that freebie, he insists on paying. It doesn't bother me, however.

Campers insist their tents and RVs save thousands by avoiding hotel and motel expenses, but the truth is that many KOAs and other private campgrounds charge at least forty or fifty dollars per night. When I checked into the Colter Campground for five nights, they said it would be seventeen dollars and fifty cents.

"You mean for each night?" I asked.

"No. Totally. It's three dollars and fifty cents per night."

No wonder our country is having a debt crisis.

# # #

Chapter Twenty-One

Obsession

December 10, 1974: From this day forward may you be fleet of foot, strong of body, oblivious to pain, and victorious in the achievement of your goals and objectives. – Words on a telegram received from a friend before the Joe Steele, Rocket City Marathon, Huntsville, Alabama, in 1978, my first marathon

From 1974 until my hip-replacement surgery in 2008, my mania for record keeping revealed that my preoccupation with running undoubtedly qualified as an obsession. My log books for those years show my mornings starting at 5:00 a.m., with daily training runs of six and ten miles and weekend runs of twenty miles. Fifty or more weekly miles had been recorded yearly during that period. Nineteen eighty-three had the most miles: 2,810.5. I used a stopwatch on every run and was spending eight to ten hours a week, four hundred hours a year, on the run. One wonders if I had devoted that time to a musical instrument, a foreign language, or any other field of concentration what I might have achieved.

    Road races were the carrot that motivated much of my training. Improvements in racing performance prompted me to train harder and smarter. I devoured books and magazines on training methods and techniques. Terms such as "fartlek" (Swedish for "speed play"), "intervals," and "LSD (long slow distance)" became part of my vocabulary. I tried any foods or beverages rumored to have racing benefits. I had shoes for training and shoes for racing, each designed for specific distances. My collection of timing devices, watches, and handheld timers filled a desk drawer. Medical remedies to treat overuse aches and pains made our bathroom look like a MASH field hospital.

So, how did I do? Was I fleet of foot? Did I qualify for the Olympics? Hardly. I didn't get obsessed with the sport until I was over forty years old. Race winners were all much younger, college athletes or in their early twenties. However, age-group finishes, such as first, second, or third in the forty-to-forty-five age group won awards, and I found I generally placed in the top 10 percent of my age group. In a race of three hundred runners, if there happened to be thirty in my age group, I usually came in third or fourth. Not great, but I collected a fair share of ribbons and trophies, and I felt athletic. Becoming a year older, particularly when you moved into the next age group, as when turning forty-five, was reason for joyful celebration, because you then became one of the younger guys in the next age group.

Racing was exciting and fun. Events always drew large crowds, and for nominal entrance fees, you received race souvenirs, like T-shirts, and all the Gatorade you could drink or bananas you could eat. Almost every weekend some organization held a race somewhere in the state. Because I had an airplane, one or two of my sons, or a friend, and I would fly to that city, borrow a vehicle or rent an airport car, and run in the local race.

Plymouth, Indiana, held an annual Blueberry Stomp each fall during its Blueberry Festival. The first quarter mile of the 15K (9.3 mile) race had blueberries spread across the roadway for runners to stomp. Cincinnati held frequent 10K races around its Lunken Airport. No need for a car when landing there. Muncie, Indiana, had a number of races on its Ball State campus, and that became a favorite fly-away destination.

Whenever I traveled for business, I always researched local road race activity. When attending a convention in Atlanta, I found Georgia Tech was holding a three-mile race on its campus. Georgia Tech, being an engineering school, rather than call their event a "three-mile race" or a "5K," called it a "π"-mile race," and the course measured 3.14159 miles long.

My runs throughout our community soon became a familiar sight about town and brought me a certain amount of celebrity. The Columbus Area Chamber of Commerce held a gala dinner to celebrate its seventy-fifth anniversary. Entertainment and skits were performed as part of the evening's festivities. One group performed a song they had written based on the tune, "Dearie, Do You Remember?" Their lyrics included, "Dearie, Do you remember, Columbus High was only one, / Before Jim Stark had even begun to run."

Because of my running reputation, for several years I was asked to organize fitness runs for conventions of the National Association of Electrical Distributors. That same awareness prompted a neighbor and chief financial officer of Arvin Industries to ask me to organize a morning fitness jog for Senator Richard Lugar, who was spending the night with him prior to a speech Lugar was giving. Word got out about the morning run, and I learned thirty to forty people planned to join us that morning. I had T-shirts printed that read, "I ran around in Columbus, Indiana, with Dick Lugar."

Senator Lugar always remembered that morning, and if I saw him at an event, he always said hi. One time I wrote a short article for the Indiana Road Runners newsletter, and Lugar sent a note saying he enjoyed the article.

Athletes run in my family, Chris and Dad, about 1978

Two of my sons, Chris and Brian, ran on the high school cross-country team, four years apart. Their coach held a weeklong cross-country camp each August before the school year. I attended the camp two of those years to help as one of the adult chaperones but also to train along with the kids. One of the events at camp was a "Flashlight Relay," run in the dark with only flashlights to illuminate the paved roadway. I was on one of the relay teams. Some years before, I began wearing dog tags around my neck for identification in case I became injured some early morning on a road. The dog tags jingled as I ran. During the relay race, even though it was pitch dark, as I passed other runners hearing my jingle, they would shout out, "How's it going, Mr. Stark?"

I served as president of my Rotary Club in 1982. One of our speakers had wangled his way into giving a program by pretending to have something to do with the U.S. Olympic Team. He was a fraud who just wanted to sell small, personal-sized trampolines for exercise. He started his presentation by ridiculing jogging as a destructive, harmful exercise that ought to be banned. Of course, the entire club knew of my running activities, even though the speaker did not. When he got to his closing pitch, "Normally sells for $250, but today, just for you, a special price of $150!" I stood up, took him by the arm, led him to his chair, and said, "I've got a $30 pair of shoes and a $2 jock, and I've been doing just fine for several years now, thank you."

The funny "rest of that story" happened a week later, when my wife got a call from her father, who lived in Lafayette, Indiana. He was a Rotary member in that community and announced he had just heard a great program and as a result had purchased small trampolines for the families of each of his children.

When my son Chris and I ran 164 miles across Florida, the newspaper articles that appeared both during and after the event were followed by speaking opportunities at service clubs in town. We gave an illustrated talk using slides we had taken during training and the run itself. I would do the introduction, Chris talked about the training, and I concluded with an account of the event itself. I have to say that we got more enjoyment out of the publicity than we did out of the actual ordeal.

Indianapolis hosted the Pan American Games in 1987. A torch-carrying relay was staged, starting twenty miles south of Indianapolis. Games competitors carried the torch to the lighting ceremony at the Indianapolis 500 Speedway. Each athlete carried the torch for only a mile or two before handing it off to the next runner. Because of the recent publicity about our Florida run, Chris and I were asked to run the entire twenty-mile relay distance in support of the runners. We felt honored to have been invited to participate.

My running stopped after I had a hip replacement in 2008. Was running the reason for the surgery? I doubt it. I know many people who had hip and knee replacements who never ran. Even if it contributed to the wear and tear, I had thirty-four years of incredibly good health, great friendships, blushing recognition, and memorable experiences. My oldest son, Eric, who for years said running was a dumb exercise, caught the bug and is presently running marathons. I can now honestly say, "Athletes run in my family."

#

Chapter Twenty-Two

Adventure on the High Seas

My sailing adventures have occurred on both the East and West Coasts, in the Greek Isles, in the Great Lakes, and on several cruises in the British Virgin Islands. My wife and I had been at sea for eight or ten days at a time; however, we were rarely more than an hour from land. I yearned for more, more sea and less shore. I wanted to cross oceans, to challenge gale-force winds, and to navigate by the stars.

One way to have that kind of offshore sailing experience is to join a crew for a sailboat delivery. Yacht owners who want their boats in tropical waters during the winter months, then back in home waters during the summer season, hire delivery skippers to make the transit, and the skippers need a crew. Hundreds of sailboats are delivered to worldwide locations throughout the year.

The Internet provides a meeting place for delivery captains looking for jobs, and individuals looking for crew opportunities.

Captain Bob Bligh — not his real name — out of Naples, Florida, went a step beyond just delivering sailboats. He operated an offshore sailing school that offered inexperienced crews a comprehensive education in offshore sailing. He charged his students a nominal fee for this education, but not much more than their share of the provisioning cost for the transit. It was a good deal all around. Bob used the yacht owner's boat for a classroom and, in turn, offered the owner a discount for the delivery. He funded the discount with the tuition he charged the crew members. The crew received the delivery experience and a comprehensive education in offshore sailing. Bob's textbooks and lectures included essential topics such as:

- Trip planning with regard to clothing, food, provisioning, and routing;
- Shipboard routines relating to navigation, watch standing, weather, and currents;
- The important skills of heavy weather sailing, sail management, knot tying, anchoring and mooring;
- Foreign port procedures;
- Equipment, such as long-range radios, sail-furling systems, and engines.

I emailed Captain Bob early in 2001 and expressed my interest in signing on for a delivery he was making from Tortola in the British Virgin Islands (BVI) to Houston, Texas. The boat had been purchased out of the Caribbean charter fleet. It would be sailed out of the Caribbean, across the Gulf of Mexico, and up to Houston, where it would be put on a truck and hauled to its new owner in California. Bob didn't take just anybody; you had to have basic sailing experience. After we discussed my background, he determined I qualified. There were two crews involved, one four-man crew to take the boat from Tortola to Jamaica, and another crew to take it from Jamaica to Texas. I would be part of the second crew, making the twenty-one-day transit, and I would meet the boat in Jamaica.

The boat was a fifty-foot Beneteau, named *Painkiller*. A painkiller is a popular rum drink in the islands. It was a lovely boat, and the four crew members each had their own cabin with head and shower.

Painkiller, 50 foot Beneteau sailing sloop, 2001

The other two crewmates were Mark, who had sailed as part of the crew that sailed from the BVI to Jamaica, and Kevin, who flew in from his home in Miami. Both Mark and Kevin were in their late thirties. I was sixty-three at the time.

On first meeting, Captain Bob seemed personable enough. The infamous Captain Bligh he was not. Mark was a college history professor and in between jobs. I would describe Mark as a former hippie, laid back, and the kind of guy who is happy to kick back and let others do the work. Kevin was in real estate, in between wives, and rather inexperienced when it came to sailing. Kevin and Mark, I sensed, were there for a vacation. I, on the other hand, was enthusiastic about the educational opportunity of the experience.

Kevin and Mark (with flag)

I had taken a Coast Guard advanced navigation course the previous winter in Florida and thoroughly enjoyed refreshing my knowledge of the subject. I first learned navigation in the Navy, particularly celestial navigation, which back in the 1960s was all we had when flying over the middle of the ocean. Long Range Navigation (Loran) was just being developed, and Global Positioning Systems (GPS) were many years away. *Painkiller* had GPS, but for training purposes the crew would use only old-fashioned dead reckoning (DR) with positions refined by LOPs (lines of position). DR navigation meant advancing your position on a chart, based on course and speed adjusted for winds and current. Mark claimed he did most of the navigating on the way over from the BVI; therefore, he left all the navigating to Kevin and me.

Navigation concepts were new to Kevin, and Bob's explanations were not clear to the novice ear. Without appearing to butt in or sound like a smartass, I found I could help Kevin with his understanding. Both Kevin and Bob commented later — separately — that they appreciated my help. Bob said, "I'm just not good at explaining things to a beginner." Hmmm, he's supposed to be the teacher, right?

One of the interesting prospects of our trip was going ashore in Cuba a number of times. In 2001 American/Cuban relations did not allow Americans to spend money in Cuba. So how could we go ashore?

Cuba was almost the last major destination in the Caribbean unexplored by the cruising community. For thirty-five years, the island had been off-limits, a tantalizing Shangri-La hidden behind an iron curtain of gunboats and coastal defenses. In 2001 the curtain partially lifted, and tourists, including cruising sailors, were welcome, albeit with certain restrictions.

Under the terms of the Cuban Assets Control Regulations, issued by the U.S. government in 1963, part of the Trading with the Enemy Act, it was illegal for any U.S. citizen to engage in economic transactions with Cuba. Although the regulation effectively prohibited spending money in Cuba, the regulations did not prohibit visiting the island. Of course, it's pretty difficult to visit without spending money, particularly since one of the first requirements is to pay twenty dollars for an entry visa.

According to Captain Bob, who had visited Cuba before, U.S. customs officials seemed to accept excuses like needing to go ashore for fuel or water and didn't ask about other purchases. Bob also advised that you obviously didn't want to be wearing a Cuban T-shirt or blowing Cuban cigar smoke in the officials' faces when clearing U.S. Customs.

The first leg of our trip involved a twelve-hour transit from Jamaica to Cayo Cruz, Cuba. During that time Kevin anguished on the navigation table for three hours until I took over. From that point on, I was never relieved. Approaching the Cuban coast was tricky, because we had to negotiate a reef. I had worked out the various turns we needed to make, determined by lines of position off prominent land features. The importance of accurate navigation became emphasized when we saw the ruins of a sailboat, about our size, wrecked on the reef.

We had no plans to go ashore that first night, just anchor, have dinner, and turn in to get a good night's sleep for an early departure the following morning. While sitting in the cockpit, enjoying our sundowners (rum and Cokes), we saw a snorkel break the surface of the water, heading toward our boat. A young boy's head emerged near our anchor chain. "Come back by the swim platform and come aboard," we said, but he refused, pointing toward the still-visible shore. Instead, he climbed the anchor chain and dropped a garbage bag on the deck. The bag contained four large lobsters, potatoes, onions, and carrots. Our curb-service vendor wanted ten dollars for his offering. No one had anything smaller than a twenty, which we gave him, feeling it was still a good buy. If Mark offered no other gifts during our cruise, his command of the Spanish language proved most valuable. Through him we discovered our visitor had a delightful personality, and we had a pleasant visit. I became moved to gift our visitor with a red T-shirt that said "Hoosiers" across its front. Mark was challenged to explain the meaning of "Hoosiers," but the young man left looking pleased.

As we prepared the lobster, another snorkel was spotted heading toward our boat. Coming up the swim ladder, now in the dark, was Manuel, who had a tote containing a second offering of lobster and vegetables. We turned down Manuel's offer, since dinner was being prepared, but he was so engaging, we invited him to join us in a Jamaican rum drink. As we were talking, we saw another snorkel coming through the water. It was becoming an international trade fair. Manuel recognized the tip of the snorkel and said, "Oh, that's my cousin, Jimmez." Jimmez also came aboard, and he had a barracuda in his bag as well as a variety of fish and mangoes. We liked both young men and felt bad not making a purchase from them, so we offered two ball caps. One had a Nike swoosh, which Manuel recognized, but the other had a U.S. Open golf patch, which confused Jimmez. We tried to explain but sensed golf was not big in Cuba.

After our guests departed, a utility boat with three Cuban guards toting machine guns arrived. They spent an hour asking questions about our boat, our identity, and the purpose of our cruise. They informed us we were allowed to anchor overnight but must depart before sunrise in the morning.

After leaving Cayo Cruz, we had perfect sailing winds of thirteen knots. This gave us an excellent opportunity to try different methods of trimming the sails. Captain Bob, an experienced ocean racer, showed us different ways of adjusting the sheets, traveler, halyards, boomvang, and fairleads to obtain maximum speed. I learned more in that one day about sail shape, sail twist, boat heel, and rudder position than I had learned in all my previous years of sailing.

As we sailed, Captain Bob became more demanding and insistent that we do things properly. I felt this was appropriate behavior for a skipper. Mark and Kevin, however, complained that the SOB Bob insisted on strict course adherence and tight sails. Why be such a tight-ass? they wanted to know. That attitude became reflected in their unwillingness to perform certain tasks or make fine-tuning adjustments to the boat's trim. Almost correct was good enough for Mark and Kevin.

We planned a two-night stop in the city of Cienfuegos. To get there we needed to sail up a ten-mile waterway at night with numerous twists and turns. I had meticulously plotted all the turns using marker buoys and points of land shown on the chart. Bob was at the helm, and we discovered that many of the marker buoys had been removed. Apparently, Cuban officials wanted to make it difficult for anyone to leave Cuba. The captain's piloting of *Painkiller* up that waterway was masterful. I was impressed and told him so, which I could tell pleased him.

Once we secured the boat in the marina, soldiers toting machine guns came aboard and put us through a rather stern inquisition. It looked serious to me, but I nevertheless asked the captain if he thought the soldiers would mind if I took their picture. He asked and, to my surprise, the stern faces broke into broad smiles, they pushed the machine guns out of sight, and the soldiers posed proudly.

Cuban customs officials (guns under table)

The next morning we needed to accomplish a long list of tasks and procure provisions for our continuing trip. Mark and Kevin eagerly discussed going into the city to check out the *chicas* (young women), and Bob had a tour he was interested in taking. After the last ten days of close quarters with these three, I was looking forward to being alone for a while. So I announced that I'd take care of the tasks and the provisioning, and they could take off. We could meet later that night for dinner. They immediately accepted my offer.

Walking into town, I looked for a taxi or some kind of transportation to go to a money exchange to change my dollars into pesos, and then go to a grocery. I saw numerous horse-drawn carts and wagons on the streets. Most of the autos were European, except for a few 1960s American cars. People I passed smiled and waved in a friendly manner, but hardly anyone spoke English. Down the street I saw a horse-drawn trolley with pedestrians climbing aboard. I surmised this must be public transportation. I inquired in sign language and learned the trolley was free, but for the use of Cubans only.

Seeing a group of men loitering around in front of a garage, I discovered one of them spoke a bit of English and seemed to know someone with a car who could drive me on my errands. "Si, si," he said, and off he ran. I hoped he went for a driver, and I waited for twenty minutes.

Finally, a small, late-model foreign car arrived, and I was introduced to Miguel. He spoke no English. My interpreter told him I wanted to go to a money exchange, then to a food store, then back to the marina.

Miguel was nice looking, late thirties, and neatly dressed. I didn't know for sure he understood where I wanted to go, so I held up a fistful of money. He nodded his head: "Si, si."

We drove to a bank-looking building that was surrounded by people. Oh my, I thought, if they are waiting to get in, I'll be there all day. "No, no. Go right in," Miguel gestured. I learned later that the people outside wanted American dollars and were hoping to intercept tourists before they went into the bank.

I wanted to exchange two hundred U.S. dollars for pesos. The teller pushed the key on a machine, and it started spitting out pesos. Soon I had a stack of money two inches high in front of me. "My gosh," I exclaimed. "Is that all mine?"

"No," the teller replied, "that's only half of it."

"I need bigger bills. I'd never get that in my pockets."

The teller took the pesos back, keyed in a larger denomination, and pushed the button. This time the stack for the entire two hundred dollars' worth was only one inch high.

Miguel waited for me until I finished and seemed to understand that our next stop was a grocery. What a nice guy he was. Something about him clicked with me. We spoke no common language, but a smile, a nod, or a chuckle at something passing on the street created a bond.

The grocery was impressive: wide aisles, new grocery carts, and beautiful displays of food. The shoppers were not the same people I saw on the streets. These were tourists, a few Americans, but mostly Europeans, Asians, and Canadians.

Miguel became a great help. In order to find the next item on my list, I'd act out a pantomime. If I wanted apples, I acted out picking one from a tree, and off he'd run and come back with what he thought I wanted. When I acted out a chicken laying an egg, he about doubled over laughing.

Soon our cart spilled over with our selections, and I pushed it to the checkout person. "One hundred and thirty-two dollars," she said.

"Yes, but how many pesos?" I asked.

"No pesos, only dollars here. This is a dollar store."

The Cuban government was trying to attract American tourists in spite of U.S. restrictions. It had created dollar stores, where only dollars could be spent. The stores had better-quality goods than other stores, and that's why there were so many people at the money exchange trying to acquire dollars.

Miguel kept hitting his head with the flat of his hand. Yes, it was upsetting. Through sign language, Miguel suggested we push the cart to the side, return to the money exchange to get the pesos changed back, then return for the groceries. Fortunately, for just such emergencies I keep a hundred-dollar bill and a fifty-dollar traveler's check folded in a secret compartment in my wallet. We were able to take the provisions with us.

Before we left, I asked two British tourists who spoke Spanish to tell Miguel that on the way to the money exchange I'd like to buy some postcards.

I didn't think Miguel understood, because he took me to a post office, but then he pointed out a small retail shop in the post office that sold postcards. I purchased a bunch using just a few of the bills from my one-inch stack of pesos.

When we returned to the money exchange, I again went right in and pulled out my pile of pesos. "So sorry. It is not possible. You are only able to change dollars into pesos in Cuba, not the other way."

Miguel looked stunned when I told him. He had become "mi amigo," and I felt a genuine friendship with this compassionate fellow. Okay, so now I have two hundred dollars' worth of pesos as a souvenir of my visit to Cuba.

Back at the marina after giving the provisions to a dockhand who put them into a wheelbarrow, I pondered how much to pay Miguel for his services. The cab ride in Jamaica from the airport to the boat, a forty-five-minute ride, cost thirty dollars. So, taking out a piece of paper, I wrote, "$40 U.S. = 880 Pesos," then started counting out fifty-peso notes. That must have been more than Miguel earned in a week, because he kept pumping my hand and calling me "mi amigo." I don't care if I did overpay him. He was worth it.

After putting away the food and doing my laundry in a bucket on the stern, I started writing postcards, twenty-seven of them. I thought my friends might enjoy seeing a postcard from Cuba, but I wondered if the U.S. Post Office would have a problem with it. After all, U.S. citizens are not allowed to spend money in Cuba, so how could I buy postcards? Then I cleverly came up with this message:

Dear John,
Your friend Jim Stark asked me to send this card since U.S. citizens are not able to make purchases in Cuba. He has completed 350 miles of a 1,200-mile sailboat delivery from Jamaica to Galveston, Texas. He looks wonderful; tan, lean, and very muscular. He hopes to stay that way and see you again back in the U.S.
    Adios,
    Santiago

Santiago is the Spanish equivalent of James. Even though the note on each card was brief, it took over an hour to write all twenty-seven cards. But I thoroughly enjoyed myself, sitting there quietly in the cockpit with the activity of the Cuban harbor going on about me.

Captain Bob returned to the boat about four-thirty. I told him about my friend Miguel and our money-changing experience. Bob said, "You realize you will never be able to change that money back into dollars once you leave Cuba?" Yes, I did, but didn't see any way around it.

Bob had business to take care of at the marina office and asked to take my pesos with him. I had seen a hotel outside the marina and wondered if they could mail my postcards, which already had postage attached.

When leaving the marina, from behind me I heard, "Mi amigo, mi amigo." It was Miguel, and he had his wife with him. She was cute and looked like she was in her mid-twenties. Miguel had returned to try to look me up. When he saw the cards in my hand, he motioned for me to get in the car, and we drove to the post office to mail the cards.

Returning to the marina, I invited Miguel and his wife out to see the boat. When we reached the pier, however, a security guard stopped us. "No Cubans allowed on the boats!" We begged and pleaded, without success. With heads bowed, we turned to leave but ran into another security guard; this one outranked the first and allowed Miguel and his wife a five-minute tour. Going aboard, I gave them a Red Stripe beer and showed them around.

Just then the captain returned, and I introduced my Cuban friends. The captain had good news; he was able to change my pesos back into dollars. That was good news, and I was pleased that Miguel heard it.

Saying goodbye to Miguel, we gave each other a big hug. He was, without question, the highlight of my Cuban visit. My only regret is that I didn't take his picture.

We departed Cienfuegos early the following morning and made our way along the southern coast of Cuba toward the Gulf of Mexico. Sailing about fifty miles offshore, even at that distance, Cuba is surrounded by reefs. Sometimes we gave a wide berth, and sometimes we carefully picked our way through. Precise navigation was of utmost importance.

There were days when Captain Bob seemed testier than usual. I tried to stay clear of him during those times, but Kevin and Mark became infuriated. There was one incident that really set the captain off, and for good reason.

Cutting through one of the reefs required an accurate heading. Kevin was on the nav table and did the calculation. Magnetic variation is the difference between true north, the top of the Earth, and what a magnetic compass reads. There are places in the world where they are equal, but other places where the variation is as much as eighteen degrees. All charts are laid out with reference to true north. For example, if you plot a course on a chart and it measures 270 degrees, you would note the variation (shown on the chart) and adjust the boat's magnetic heading. If the variation is five degrees west, you add the variation and sail a magnetic heading of 275 degrees. Variation is west in some areas and east in others. Knowing whether to add or subtract the variation is critical, and Kevin did it wrong. We headed through that reef ten degrees off course.

I was at the helm and should have checked Kevin's calculations. The slot we were sneaking through was narrow. We were under diesel power at the time. Suddenly, I saw a splash of whitewater one hundred yards directly in front of us. I put the engine in reverse to get the boat stopped quickly. That brought the captain on deck. "What the fuck is going on?" he wanted to know. Then seeing the reef directly ahead, he disappeared below to check the GPS. "We're two hundred yards left of course," he roared. "Turn starboard 360 degrees and get us the fuck out of here!"

"Aye, captain."

Once back on course, the captain calculated the correct heading, and disaster averted. I was held responsible, and Kevin's name never came up in the reprimand. Of course, if this $500,000 boat had been wrecked, it was the captain's ass on the line.

Days with winds in excess of twenty knots meant spirited sailing. With the boat heeled over to thirty degrees and the bow plowing through eight-foot seas, water cascaded down the deck. How exciting! One of the items on my wish list for this trip was to experience gale-force winds of forty knots. "What? Are you crazy?" people ask. "Those are serious, threatening conditions." It would not be fun, no, but a veteran sailor ought to have experienced those conditions.

One of the fun sights was dolphins playing in the bow wave. We'd be sailing along at a brisk seven knots when, out of nowhere, suddenly twenty dolphins shot across the bow, leapt into the air, and blew spray out of their blowholes. I always assumed it was the pressure wave created by the boat's forward motion that attracted them, but the captain, who knows everything, said, "No, they're just a bunch of show-offs who enjoy leading a parade. The chirping you hear is their signal to others, 'Hey, there's a boat, come join the fun!'"

The other unique experience had flying fish landing on the deck at night. I told the captain I thought it was cool, and he said, "Pretty ordinary, actually. When out here on open water, if you don't get hit in the face by a flying fish some night, it's unusual."

For the record, I never got hit but spent every morning flinging dead, slimy flying fish off the deck into the sea.

The western tip of Cuba at Maria la Gorda was our last night at anchor before heading across the Gulf. From there it would be 710 nautical miles (817 statute miles) to Galveston Bay. The heading was 310 degrees, and we anticipated it taking 112 hours. It would be around-the-clock sailing for four and a half days. That was an experience I looked forward to. The captain said, "You'll change your tune. It won't be fun, and you'll be exhausted before it's over."

We would be standing watches. During the daylight hours, 7:00 a.m. to 9:00 p.m., we operated in a normal manner; all hands shared the work, and we ate one hot meal together. The remaining ten nighttime hours, we stood three-and-a-half-hour watches. Kevin and Mark were paired, and I was teamed with Captain Bob.

I was surprised that Kevin didn't want to be matched with Mark. Kevin actually had more capability than he gave himself credit for, but the Gulf was a busy place. You're almost never out of sight of huge transiting tankers either heading for or departing the ports of Galveston or New Orleans. At night one must be alert to make heading changes to avoid a collision. Kevin felt he needed a reliable watch partner, and Mark's "don't give a shit" attitude bothered him. I didn't blame him.

Tankers in the Gulf of Mexico

Winds in the Gulf were strong, although from the wrong direction. Sailboats can tack back and forth and go in any direction, but we needed to maintain a course of 310 degrees and a speed of seven knots to stay on schedule. Our best choice was to run under diesel power — referred to by sailors as the tongue-in-cheek "iron jenny." We had plenty of fuel and could maintain seven knots easily. That also gave the helmsman more flexibility in dealing with traffic.

Because falling overboard at night was a life-threatening event, all of us wore safety harnesses. Whenever going on deck you needed to attach one end of the tether to part of the boat.

I loved the routine. I had no trouble falling asleep at the end of my watch and, surprisingly, awoke without prompting three hours later, ready for my next turn. Sitting in the cockpit at 2:00 a.m. watching the Big Dipper slowly rotate around the North Star was magical. Captain Bob worked on his computer below deck during most of our watches, so I was left alone with my thoughts and to deal with the tankers.

Day is done

The tankers stayed within highways across the Gulf called "fairways." The ships received a clearance to be there. The fairways are clearly indicated on the charts, and although we were allowed to cross them, we were not allowed to sail in them. Fairways are the most direct routes to the ports, so we sailed just outside the edge of them. Collisions shouldn't be a problem, but there are no boundary lines painted on the water, so vigilance was important.

In the black of night, you first see a tanker's running lights. The helmsman takes a bearing on the lights every few minutes. If the bearing changes, you can relax because the ships will miss each other. If the bearing remains unchanged, you are on a collision course and should alter heading.

One morning I came on deck at 6:00 a.m. during the last hour of Kevin and Mark's watch. Kevin was concerned about a ship directly ahead of us that had not changed its bearing. He kept asking Mark what he ought to do, and Mark, in his usual apathetic manner, said, "Do whatever you want to do."

That put Kevin over the edge, and he shot back, "Come on, damn it, we're supposed to be working together here!" I didn't blame Kevin for being upset.

"Ask Stark," was Mark's flip suggestion.

I contacted the ship by VHF radio and found out the ship was stopped, waiting for orders. "Kevin, why don't we maintain our heading until one mile from the ship, then make a heading change to have the tanker slip by our port side?"

"Sounds good."

Mark sat in the cockpit the whole time and watched passively.

During the day, all crew members were busy, tending the boat. Such things as changing the oil in the diesel engine and reversing the jib sheets — lines used to control the foresail — are easily accomplished tied to a dock, but when the boat is heeled over thirty degrees and pounding into six-foot seas, it adds somewhat to the challenge.

The 310-degree course had to be adjusted periodically to keep Galveston directly on our bow. Kevin couldn't understand this. I explained that charts and maps are flat, but the Earth is round. Laying out a straight line on a chart to your destination does not take into account the curvature of the earth. Great circle routes, such as those I used in the Navy when flying across the ocean, required adjusted headings to achieve the shortest distance.

One hundred miles from Galveston there are literally thousands of oil platforms. You could not sail a straight line without hitting one of them. In the darkness they look like brightly lit Christmas trees rising a couple hundred feet in the air. They're huge, with helicopter pads, large hoisting derricks, and living accommodations for scores of workers. Fascinated by them at first, we soon became accustomed to their presence, as though they were buildings along the highway.

Galveston is four hours from our final destination, Clear Lake, which is inland and just south of Houston. There is a waterway from Galveston to Clear Lake. Before sailing up the waterway, we first had to clear Customs and Immigration at Galveston. We anticipated the stop to take a couple hours to complete the requirements. We planned to acknowledge our visits to Cuba, but we felt a wee touch of anxiety that we might be challenged in some manner. Each of us took the precaution of stuffing the few items we'd purchased, T-shirts, flags, and postcards, into the bottoms of our dirty laundry bags.

The captain radioed Customs, and we were cleared without stopping. Great news! Next, he contacted the immigration authorities. They were not so agreeable and said we needed to stop to provide a crew list. The captain asked if it was possible to give the crew list to authorities when we arrived at Clear Lake. The Galveston immigration authorities said they would check with headquarters. In the meantime, we continued up the waterway toward Clear Lake. Impatient to get an answer from Immigration, the captain started badgering them. His insistence became more and more demanding with each radio call. We were two hours up the waterway when Immigration said, no, we had to turn around, return to Galveston, dock the boat, take a taxi downtown, and produce the crew list.

Kevin and Mark went ballistic. They wanted off the boat! Instead they were now looking at another half-day of having to put up with Captain Bob. When we docked at Galveston, and as I secured the stern lines, up from below came Mark, carrying his sea bag. The captain said, "Oh, I see Mark is going to leave us." Next to appear on deck was Kevin, also with his bag. "And I see Kevin is going to leave us as well. Are you going too, Jim?"

"No, sir. I'm here for the duration."

Before the captain and I got into the taxi, Kevin and Mark pulled me aside to say good-bye. I really liked Kevin, although I sometimes found his behavior rather immature. Kevin said, "Sorry to do this to you, Jim. But I can't stand another minute of that SOB!"

Although Mark and I had shared some pleasant conversations, I didn't have a lot of respect for him. He surprised me when he said he hoped he would become more like me when older. I found his remark strange, since we were so different from each other.

In the cab, the captain said little about our departed shipmates other than it was a reflection of their characters. The immigration visit took five minutes and we were soon back to the boat. Now instead of three people handling the dock lines there was one — me. Busy, busy!

I thought, naively, that my loyalty to the captain might be rewarded by him going easy on the workload. Wrong! When I asked if there was something I ought to be doing during our four-hour transit up the ditch to Clear Lake, Captain Bob said, "Well, all four heads need to be cleaned, the food needs to be taken out of the refrigerator and freezer, and they need to be scrubbed down.

We tied up at the dock at Clear Lake at 9:00 p.m., then took another hour to wash down the deck and dispose of the trash. Before departing for the airport to make arrangements to fly home, I bought Captain Bob a drink at the marina bar. I had no regrets regarding the trip and told Bob I appreciated the wealth of knowledge I had learned at his knee. Bob, not given to compliments, said some nice things about my role as a student, that he felt I could make a trip like the one we had just completed on my own, and we parted as mutually respected teacher and pupil.

My days aboard *Painkiller* will be long remembered. While standing one of the 2:00 a.m. watches, I entertained myself by composing a poem.

Painkiller's 1,300-Mile Voyage

By Jim Stark

We joined her in Jamaica, at Port Antonio,
A sailboat named *Painkiller*, a fifty Beneteau.
She sailed in from the Virgins, early in the week.
Crew one had just departed, as homeports they did seek.

Captain Bob, Mark, and Kevin, and I made up crew two.
We prepared the boat for sailing, three days of tasks to do.
Provisions, storing, orientation — getting ready to set sail.
Soon it was off to Communist Cuba, and Texas without fail.

We learn navigation and shoot lots of LOPs,
Can tie our knots with eyes closed, as simple as you please,
Shape, twist, and trim are watchwords to make our boat's sails right.
We make it second nature for the darkness of the night.

Kevin is our food chef, he spices up the meals.
Bob offers subtle comments on the gas pains that he feels.
Mark does grilled chicken and makes the potatoes mashed.
The chicken is real tasty — the potatoes, they get trashed.

Ham is one of the staples we have many ways,
On bread, in pasta, eggs, and salad — many times in 20 days.
I wash the dishes daily, it's teamwork while at sea,
They get the food all ready, and I do all KP.

Cienfuegos in south Cuba, a city on the coast,
Is a 20-hour shore leave and serves as our port host,
Mark and Kevin look for *chicas*, Bob a tour attends,
I buy more provisions, and meet a new best friend.

Miguel is my cab driver, we speak no common word,
But he is "mi amigo," and with gestures we are heard.
He brings his wife back later, to visit our small ship.
Do you think "mi amigo" was influenced by my tip?

Coastal Nav is demanding, giving stress and perhaps some grief.
A variation error can put a boat upon the reef.
Kevin's classic question, after plotting out a plan,
After peeking out the porthole, asks, "What's that, land?"

Bob's a high-tech sailor, has website and side-band too,
Gets weather fax by laptop and e-mails for the crew.
His tiny digital camera snaps the crew while out at sea;
View them on his homepage, and take the cruise for free.

Many anchorages later, we enter Gulf of Mexico,
It is perfect sailing weather, and seven knots we go.
Hours of around-clock sailing, 700 miles north by west,
Sleepy 4 a.m. watches, with rigs and ships to test.

Finally, Galveston Bay marks the near-end of the trip.
A wonderful sailing adventure, a classroom on a ship.
Good guys, good fun, good learning — a wealth of skills acquired,
Three cheers for offshore sailing — and future trips aspired.

Painkiller Kevin, Mark, and Chicas

#

Chapter Twenty-Three

Canyon Conversations and Bambi

I first visited Grand Canyon in 1986 when my oldest son, Eric, and I barnstormed the West in our airplane. The park was not nearly as crowded back then as it was this time. Fortunately, I'd made reservations for a campsite ahead of time. The seven-thousand-foot rim altitude meant temperatures were much more comfortable than the triple-digit temps of Phoenix, Arizona. However, the day's travels wrung me out and I crawled into my tent early that evening to rest up for the hike down Bright Angel Trail the next day.

One question I am often asked about my solo adventures is, "Don't you get lonely?"

My answer: "Heck no. I meet people all the time and engage in fascinating conversations."

Like the woman I met at the trailhead the next morning. She wanted to know where they kept the mules. "I'm supposed to meet my husband there to join a mule train for a ride down the canyon," she said. "I really don't want to go. Looks dangerous to me. Why, what if those beasts slipped? That'd be it. Thousands of feet over the edge. This whole canyon thing is my husband's idea. Believes he's a big outdoorsman. Shoot, his only exercise is walking to the mailbox. But I said, I'd do it, and I'll tell you why. Next year we're going to Hawaii … by God!"

Mules, Grand Canyon, 2003

    The guy I shared a shady rock with while on the hike lost his job the year before. "No one wants to hire a fifty-year-old accountant, it seems. Then my father died. Left me a little money and a new Cadillac he just purchased. Man, he was proud of that car. Couldn't afford it, really. Said the hell with it and bought it anyway. So I decided in honor of Dad, I'd get in that car and take a little trip. Even go to the college reunion I didn't think I'd attend. Been gone three months. No plans. Just got in the big Caddy and drove."

    One man I walked near, one about my age, discovered I traveled by motorcycle and camped in a tent. "Did that once," he said. "Had a 500cc Triumph. Rode the whole West Coast. Just a kid at the time. Considered going to law school. My brother was a lawyer, made good money. But he got in trouble for bribing a judge. Disbarred. I became a teacher and a coach. My wife was a teacher. Between the two of us, we did okay. Not a lot of money, but okay. Bought us a little RV after the kids grew up. She's back at the campground now. Maybe see you later."

The hike back up Bright Angel Trail took twice as long as going down, but I'd expected that. Once I got back to the rim, I found a laundromat and washed some clothes. Not so exciting, but I was out of clean underwear. While the clothes were in the washer, I rented a shower. Yeah, no kidding, you put quarters in the shower faucet.

Hiking Grand Canyon, 2003

The dryer didn't completely dry the underwear, but I took them back to my campsite and spread the damp articles on the picnic table and weighted them down with stones so they wouldn't blow away in the breeze. Next, I headed to the park cafeteria, where I nursed a sandwich for an hour and a half while I charged my computer, phone, and camera from a wall outlet. When I returned to the campsite, I found big footprints on my drying underwear. Crows had walked all over the table, leaving sandy prints as large as your hand. Even more interesting, I had left a polystyrene washbasin of water on the table. The crows had pecked a hole in the side of the basin, flooding the table, soaking my underwear, and ruining the basin. Such is life in the wild.

After leaving Grand Canyon, I spent the next night in Durango, Colorado, working my way toward Boulder to visit my college roommate. We'd both played on the North Central College football team. Jim's a big guy, well over six feet, who played defensive end. His nickname was Bambi. So how does a big, rough, tough football player get the name Bambi? In high school all Jim's teammates taped fearsome messages to the front of their helmets, like, "Killer," "Smash," "Death," etc. Jim wanted to do the same and wrote "Bam Bam" on his helmet, but when the coach saw the words, he thought it read "Bambi." The name stuck. When Jim went to college, he hoped the nickname might be left behind, but a former high school friend spotted him on campus and spread the word.

Author and Bambi, (Jim Hickman) 2003

Bambi's nose had been broken several times in high school games. Those injuries left a rather prominent, crooked proboscis. Being the kind, sensitive roommate I was, I never tired of pointing out roomie's big nose. I was also a lineman, playing both offensive and defensive guard, but never suffered lost teeth or visible scars. During a game in our junior year, I took a huge smack in the face that sent my helmet sailing across the field. Dazed, I stumbled back to the huddle. Bambi took one look at me, pointed, and started laughing. My nose was smashed flat against the side of my cheek. Our quarterback called timeout and sent me to our team doctor on the sidelines.

I'll never forget his treatment. He placed the palms of both hands on either side of my nose and gave a mighty jerk, realigning my nose to its original position. I fell to my knees and about passed out it was so painful. But the nose healed straight, never showing its temporary displacement.

In our senior year, Bambi had his nose fixed. The school's insurance covered the operation, and he figured he might as well take advantage of the opportunity. He scheduled the surgery in a Chicago hospital during our spring break that year. Another friend and I took our girlfriends and went skiing in Winter Park, Colorado, that week. When we got back from the ski trip, all four of us went to the hospital to check on Bambi. A nose job is a rather painful procedure. I understand the doctors rebreak the nose and reshape it to a more normal appearance. The four of us walked into our friend's hospital room in our white turtlenecks, sporting glowing tans from the sun and snow. We were shocked at the sight of Bambi lying there with two black eyes and a swollen face. At first we were stunned. He was almost unrecognizable. But then someone smiled. That led to a giggle. And soon the four of us couldn't contain our laughter. Bambi said, "Okay, you snowbirds. Get the hell out of here!"

Bambi was an even better swimmer than he was a football player. He had a national ranking in the backstroke during his college years. North Central, even though only a thousand students, had an outstanding swim team and used to compete against the big boys, including Indiana University.

And that was during IU's coach Doc Councilman's glory years. North Central sent three swimmers to the Olympics for the 1960 competition.

So what was Jim Hickman, alias Bambi, going to look like after forty years? I wondered. How did he expect Jim Stark to appear? Jim greeted me at the end of his drive as I rolled to a stop. I pulled off my helmet. We looked at each other smiling, then burst out laughing. Who would have guessed that forty years earlier, two young guys lying in their bunks at NCC, full of ambition and assumptions about their future lives, would this day be hugging each other as sixty-four-year-old senior citizens? Who would have imagined all the twists, turns, and adventures life held for us? A gypsy fortuneteller reading the details of our lives in her crystal ball would not have been believed.

I accepted Jim's invitation to pull off my leathers, change into swim trunks, and soak in his backyard hot tub. The swirling waters provided a miraculous cure for my sore butt and aching shoulders. Afterward, the three of us, Jim, Shirley (Jim's wife), and I sat on the patio, reminisced, and filled in the forty years of career, exploits, and history. Jim got his law degree after North Central and practiced law out of a Chicago office. When I asked if he specialized in a particular area of law, he chuckled and said, "You know what I tell prospective clients when they ask that question? I tell them I specialize in all categories of law. One client didn't get my exaggeration and said I was exactly the kind of lawyer he was looking for."

Shirley was a good sport about hearing the rehash of the North Central stories: the beer parties, old girlfriends, naughty athletic club initiations, favorite professors, least favorite professors, part-time jobs, ad nauseam. To my surprise, most of the tales I had been telling over the years were just as Bambi remembered them. It's interesting that whenever I think back to our college days, Jim Hickman becomes Bambi in my thoughts.

One of the stories we both recalled in vivid detail involved a night football game played during a violent thunderstorm. At the moment the ball was snapped from center and players crashed together making their blocks, lightning flashed, and all the lights on the field went out. Most of us remained still on the ground or wherever we were at that instant. But out of the darkness we heard our left tackle, a large but not very bright individual, yelling in anguish, "I'm blind, I'm blind."

Bambi had a perspective on our coach that I had never considered. Our coach, Jesse Vail, knew a lot of football but had difficulty keeping his emotions under control during games. The assistant coaches used to have to take over, almost tying him down on the bench. We made a lot of fun of Coach Vail. Bambi said, "Did you ever realize how much Jesse did for the minorities on our team?"

I never did, although I was aware we seemed to have a significant number of black players on the team — many more than other teams in our conference. Many of the players came from Mooseheart, an orphanage in the Chicago area, and had been given full scholarships. We once traveled to Tennessee to play Fisk University, an all-black college. Fisk held a pep rally before the game and invited our team to attend. Coach Vail was introduced, and he in turn introduced the team. When the Fisk students realized how many black players we had on our team, they stood and gave our coach a standing ovation. Their student body understood what our coach had done, even if I did not.

That evening the Hickmans took me on a tour of Boulder, including dinner at their favorite restaurant. We ate our meal on the bistro's lofty deck overlooking the city of Boulder spread out before us. Later, back on the patio with after-dinner drinks in hand, the stories and reminiscing resumed.

It was a good visit with my former roommate. Jim is an avid fly fisherman. Michele and I tried to learn the sport the previous year when hiring a guide in Michigan to teach us the skills. We thoroughly enjoyed the experience, but the payback for our equipment investment was one rather small, ill-fated trout. "You want to catch fish?" Jim asked. "I'll show you how to do that. You and Michele fly out here, and we'll go fishing. I promise you'll catch fish."

And we did three weeks later. Just as Jim had promised, Michele and I caught fish!

My final week out West included a number of highlights, such as camping at Rocky Mountain National Park and visiting wild and woolly Dodge City, Kansas, but soon it was time to head home. Once again I marveled at my good luck with no mishaps. I try not to boast about my good fortune, for fear I might jinx the circumstances. And sure enough, just as I was patting myself on the back, misfortune struck.

I tell myself the purpose of my travels is to document my experiences in narrative and photographs. At least that is the unselfish excuse I use for doing these travels. Therefore, every night at my campsite, I spend a good deal of time downloading the pictures I've taken and writing up the day's happenings. Imagine my concern when, with only a couple days left on this trip, my computer wouldn't turn on. Help! Was it kaput?

Chances were that an internal component had failed and the hard drive was okay, but I had no way of knowing. I had sent many of my journal entries home in emails, so that documentation was on other computers, but still, I worried. This all happened on a Thursday night, nine hundred miles from home. If I could cover those nine hundred miles in two days, I might be home Saturday afternoon, early enough to get to a computer shop, and have the hard drive pulled before the weekend. There was no real need for the rush, other than peace of mind. I urgently wanted to know that my pictures and documents were safe, all in one place.

Riding long miles on a motorcycle is not like driving a car. Your position on a motorcycle seat is somewhat fixed, without much wiggle room. I have cruise control, enabling me to take my right hand off the throttle, and foot pegs that allow me to stretch my legs somewhat. But the shoulders, back, and neck on an old guy let you know their protests.

The next day I rode across Kansas, covering five hundred miles, crossing the state line at 7:00 p.m. If I could get another hundred miles into Missouri, I would be within reach of Indiana on the final day. I wasn't on an interstate, and the numerous small towns I crossed slowed my average speed. Finally, I'd had it and needed to stop. The next motel would be the end of my day. Nope, nothing was found in Collins. Hermitage, being a county seat, surely would have a motel. Nope. The guy at the gas station said, "Only place I know is Lake Ozark, fifty miles down the road." Oh, boy. I was tired and that would be another hour!

Somehow I made it, six hundred miles. A new record for me. Lest I become too proud of myself, there is a motorcycle award called the Iron Butt Club. Membership is awarded to anyone riding a thousand miles in twenty-four hours. No thank you.

With only three hundred miles to cover the final day, I was able to pull into PC Max in Indiana at 4:00 p.m. The technicians completed the diagnosis quickly and within minutes handed me a disk with the entire contents of the hard drive. Even more good news; the compact laptop I carried on these trips was going to be easily fixed.

What a great trip! I covered 4,800 miles in twenty-one days, with only two of those days in the rain. What sights I saw and history I had learned. What a great adventure. America the beautiful, for spacious skies, for amber waves of grain ... America, America, God has shed his grace on thee.

And thank you, Guardian Angel, for keeping the shiny side up! What a lucky guy I am.

###

Chapter Twenty-Four

Return to Paradise

Two years after our first bare-boating charter with friends in the British Virgin Islands in 1996, just the two of us, Michele and I, returned for an eight-day cruise. Bare-boating means acting as your own captain and crew, and this was as much a test of our ability as a couple to handle a large offshore sailboat as it was a vacation. Michele and I intended for sailing to be a big part of our retirement years, and it would be wise to confirm that the two of us could handle the inevitable challenges that occur on a large boat. Two people can easily sail a fifty- or sixty-foot boat as long as diesel engines, furling sails, power winches, and electronic communication and navigation systems operate properly. It's when mechanical and electronic devices act up that extra hands and manpower become desirable.

Two of those snafus occurred on the first day of our cruise. Lazy jacks had been added to all the Moorings boats since our last charter in the BVI. This is a system of lines attached to the mainsail's boom and mast that gather in the sail when it's lowered, rather than have it pile up on the deck. It's a good system; however, we had been warned that lazy jacks sometimes cause the sails' battens to get caught in the lines when raising the mainsail. Just lower the sail and clear the entanglement, we were told.

Michele at the helm, 1998

That first afternoon we had stopped at one of the islands for lunch. When we left and raised the sail, it became fouled. We didn't notice it immediately, not until after shutting down the diesel engine. Then we noticed the shape of the sail wasn't right. We restarted the engine and lowered the sail, but Michele wasn't able to free the batten tangled in the lazy jack. I had her take the helm to keep the boat headed into the wind as I went on deck to wrestle the sail free. Raising the sail, it again became caught. It took three attempts to finally get the sail up without its being fouled. Later, we figured out the trick to get the sail raised without getting it snarled.

Securing the boat in the evening became the next challenge. Because of numerous reefs throughout the BVI, sailing after dark is prohibited by the charter companies. The BVI national park system has placed a number of mooring buoys throughout its inlets, bays, and harbors. When securing for the night, it is an easy process to sail up to a securely anchored mooring buoy, hook on a bowline, and break out the sundowners for end-of-the-day libations.

When buoys are not available, one must set an anchor. Most bays and harbors in the BVI are deep; therefore, to get a good anchor set, you need to be in shallow water close to shore. Not too close, of course, because that's where the rocks and coral heads are found. If anchoring in thirty feet of water, the rule is to allow five times that depth (one hundred fifty feet) for scope, the length of anchor chain lying on the bottom from the anchor.

The 150-pound anchor and chain are raised and lowered from the boat by a powered anchor winch. After picking a spot, making sure you have enough swing to stay clear of other boats and the shore, the anchor is lowered by taking the winch out of gear but keeping control of the anchor brake so that not too much chain is released. When the anchor hits bottom, the helmsman backs the boat until the proper scope has been laid. It's a two-person, coordinated maneuver, and Michele just couldn't control the anchor brake; either too much or too little chain was being let out.

We discovered this difficulty the evening of our first night. I was at the helm, forty feet away back by the stern, trying to shout instructions and encouragement to Michele. After numerous attempts, we realized she wouldn't be able to do it, and I would have to lower the anchor. Michele was also uncomfortable backing the boat. Therefore, at the helm, I would maneuver the boat into position, then slip it into reverse. As the boat came to a stop and started slowly backing, I ran forward, released the anchor, and watched for the proper length of scope as the boat backed up. When near the set point, I'd lock the anchor winch, dash back to the helm, and bring the boat to a stop when the anchor seemed to be set in the sandy bottom.

The final step is to dive over the side, swim down, and inspect the set of the anchor. If all looked good; then, at last, it was time to breathe easy and break out the rum and Cokes.

The lesson learned was to get into a harbor early, while mooring buoys were still unclaimed. Michele excelled at snatching a buoy with the boat hook and clipping on the bowline.

Michele's lack of muscle power was more than offset by her culinary skills. The food provided by the Moorings, per our order, included breakfast and lunch fixings for every day and dinner ingredients for half the nights. The other nights we would eat at a shore restaurant. Our boat, named *Dream Aloud*, a thirty-six-foot Beneteau, had a full galley, with stove, microwave, refrigerator, and freezer. A charcoal grill hung on the rail back on the stern. We had a number of delicious meals on board while sitting in our spacious cockpit, watching a golden sun sink behind the island's mountains to the west.

Over the next few days, Michele and I revisited the favorite places we had first discovered two years before. They included Peter Island with its caves, and the Baths on Virgin Gorda.

It was especially hot and humid in the BVI that year. Natives claimed El Niño caused it. Just moving about took an effort; that explained the local natives' slow-motion saunter. On the third day, after securing our boat at the Bitter End Yacht Club in Virgin Gorda, we decided taking a taxi tour of the island's rocky mountaintops might offer some cool breezes. The taxis are pickup trucks with bench seats in the truck bed. A canvas top acts as a shade but is open on all sides.

Goats are everywhere on the island, especially around the next blind corner. "Goats don't belong to anyone," our driver told us. "If you kill a goat, it's no big deal. Cows also roam the island unattended. If a cow gets in your garden, no one owns it, but if you kill one with your car, suddenly that cow is everybody's property."

We slipped and slid across the taxi's vinyl bench seats but hung on as the driver negotiated the chuckholed, eroded roadway. While at a high point on the island, the driver pointed out a hurricane hole down on the beach. A hurricane hole is a protected pond accessible by a narrow channel of water into which boats can get out of a storm's forces of wind and sea. The hole the driver pointed out had a pond of very small dimensions. It could hold twenty boats, perhaps, if each made itself tight on either side. He had seen boats tucked into that hole on several occasions.

The Little Dix Bay Resort became the next stop on our tipsy taxi tour. The Rockefeller family owned the property at one time. It is breathtaking. If anyone offers to send you on an all-expense-paid vacation of your dreams, tell them you will take a week at Little Dix Bay during the winter. A one-bedroom suite with ocean view, in 1996, cost $10,500 per week. The taxi stopped, and we had fifteen minutes to wander the grounds. Everything was tucked back into the palms and vegetation. All areas — dining room, tea room, and lounge — are separate buildings with no walls, just open sides surrounded by foliage and flowers. It seemed strange to see a library with book shelves, leather armchairs, and sofas right at the edge of the lawn. There obviously had to be provisions for some pull-down protection from the frequent tropical showers.

We passed a British couple and asked how they were enjoying their stay. "Oh, quite nice, really. Very quiet and peaceful, actually." Typical British wild enthusiasm.

We asked the taxi to drop us off at the Mine Shaft Restaurant for lunch. We'd call him later if we needed a ride back to the boat. The restaurant sat on top of an old copper mineshaft, the highest point of the island, with a dramatic view for miles about us. A couple sat down beside us and after an opening inquiry of "Where ya from?" began a friendship that led to the couple sitting on the deck of our boat four hours later.

Tom and Barbie lived in Colorado. He worked in construction and reminded me of other rough-and-tumble contractors I had known over the years. Tom had a number of adventurous hobbies, like hang gliding, and told great stories. Barbie, ten years or so younger than Tom, grew up in a wealthy family. They were staying in a home on Virgin Gorda owned by her parents. Her dad was a surgeon, and her mother a psychiatrist who actually practiced in Virgin Gorda four months of the year. Barbie said her mother had heard someone say you had to be crazy to live in the tropics, so she thought, Hey, this would be a good place to set up a practice.

Our friends introduced us to a new island drink called bushwhackers that sounded sweet but is served frozen and is very smooth and refreshing. It's made with vodka, Bailey's Irish cream, and Kahlua. After three of those, we decided it replaced painkillers, our previous favorite island drink.

During our conversation, I told Tom and Barbie about my son Brian, who at that moment was running across the United States. It was a nine-month adventure, and although unsupported, Brian would periodically get into a library and use a computer to let his family know how he was doing. Brian's older brother, Eric, and I had flown out to Missouri in my plane several weeks earlier and followed along with him on bicycles. His mother would rendezvous with him for a time after he got to Utah.

Tom, fascinated by Brian's story, asked where he was presently. "He ought to be entering Colorado any day now," I said.

"We're going home tomorrow. If you have a chance to communicate with Brian, tell him to give me a call if he's anywhere near Aspen. I'll drive and pick him up and would love to host him for a day."

I reached Brian several days later, just as he approached Aspen, and told him to call Tom. A week or so later I heard about Brian's interesting visit, a visit that ended up in Brian's book about his trans–United States run, titled, *Getting to the Point: In a Dozen Pair of Shoes*. (Brian's cross-country run ended at Point Reyes, California.)

As Tom had promised, he picked Brian up, and they spent a fun-filled day together, which included fly-fishing. While fishing, Tom told Brian an incredible story. Tom said one day one of his longtime employees failed to show up for work, and he didn't come back to work for a week. When he did return, he asked Tom, "Don't you wonder where I was?"

Tom said, "Yes, but I figured if you wanted me to know, you'd tell me."

"Well, I do want you to know," said the worker. "You might recall me saying that my sister married a real SOB. The guy was beating on her. My brother and I told him, if you hit my sister one more time, we're going to get you. Well, he did. He beat her up again. So my brother and I killed him."

"What?" Tom exclaimed. "You killed him? My God, man, that's ... that's ... that's murder!"

"Yep. Shot him in the head. Deader than hell."

"But ... but ... what about the police? What happens when they find the body?"

"Don't think it will be a problem," the worker said. "We put the body in the back of a pickup truck and drove it up into the hills by those old abandoned mine shafts. We threw it down one of the shafts, then dropped in ten sticks of dynamite. No one's ever going to find that body!"

"But what about the police?" Tom asked. "Won't they be looking for him?"

"No, I don't think so. We had the sheriff with us at the time."

Brian's book had other amazing stories, but that one about western justice was the most interesting.

Michele and I had docked at the yacht club in Virgin Gorda for one other purpose. We planned to join a flotilla for a sail out to Anegada, the outermost island of the BVI. During the briefings back in Tortola, prior to the start of the charters, sailors are told first-time visits to Anegada are prohibited without a guide to show you the way. Then the briefer said that, although they are not allowed to, there are always a few people who try it on their own. Just know this. If you sail to Anegada on your own, and after four hours you don't see the island, you aren't going to Anegada, you're going to England!

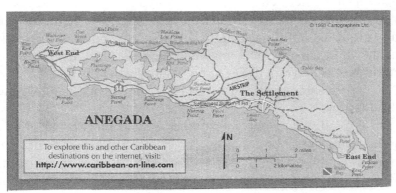

Anegada Island, BVI

Anegada is known for its fantastic snorkeling and having the biggest and most delicious lobsters in the Caribbean.

Michele and I signed up to join a flotilla of fourteen boats, shepherded by an experienced guide. The night before our departure, we met our fellow sailors at a cocktail party. We were given goodies like T-shirts and flotilla pennants for our boats, and we heard a briefing about the sail.

The magnetic heading would be 010 degrees. Anegada is surrounded by reefs. There is only one safe entrance through the reefs. It's marked by red and green buoys, but the guide boat would be anchored near the buoys to make sure everyone entered correctly. Twelve mooring buoys are available inside the harbor. Other boats would already be moored there, so no telling how many buoys would be available for the fourteen boats of the flotilla. When hearing that, I could tell tomorrow would be a sailboat race to try to get there first.

One of the interesting four-man crews we met was John and Marshall, partners in a computer-training company in Denver, and their wives. Marshall was an experienced sailor who owned a thirty-foot boat of his own. John was blind.

John could just see shadows, and you were not aware of his blindness unless watching closely; then you saw he was always touching either Marshall or his wife. His wife also helped by putting food on the plate for him. John had a wonderful sense of humor and made a number of jokes about their week so far. "This has been the most beautiful place I've ever seen." That crew also had challenges with anchoring and referred to the activity as the "A" word and forbade its being spoken.

A variety of boats made up the flotilla, most larger than our thirty-six-footer. The fifty-foot sloops, catamarans, and one J-120 all sailed faster than our smaller boat; therefore, Michele and I decided to get a jump on the pack by departing an hour before the suggested 9:30 start. It was permissible to do that, and at our departure the next morning, we saw Marshall and John's boat, and a couple of others, doing that as well.

Three of us seemed to be tied neck-and-neck almost the entire way over. Michele and I took thirty-minute turns at the helm, just to share the fun, but when it became her turn, I made minuscule adjustments to the trim of the sails, trying to coax another tenth of a knot out of our speed. We approached the entrance buoys slightly ahead, but one of the other boats had an inside position, and it would be first in the single-file entry. We were second, with the third boat close on our stern.

Once inside the bay, we could see four unclaimed buoys. The two other boats made a dash for the two buoys closest to the starboard. I saw one to the port, farther away, but it appeared to be ignored. Michele caught the mooring with our boat hook on the first pass, and we were securely moored in short order. Four other flotilla boats had beat us into the harbor and had already tied to mooring buoys.

One buoy remained unattached, with one of our flotilla mates heading toward it at a rapid speed. I recognized the two-person crew from the cocktail party the night before. He was a twenty-something, suntanned Adonis, with adequate sailing experience. She was a gorgeous, bikini-clad knockout who had never been on a sailboat in her life. Bikini babe was on the bow, boat hook in hand, while Adonis controlled the helm. The boat was going much too fast. All surrounding crews froze on their decks, watching the maneuver unfold.

Baby Doll caught the mooring buoy with the boat hook, but as the boat sped forward, the hook was yanked from her hand, still attached to the buoy. What now, Adonis?

All the surrounding observers became horrified when pretty boy slammed the boat into neutral and started stripping off his shirt as if he were going to dive overboard to retrieve the floating boat hook. All observers screamed with one voice, "NO, NO, DON'T DO IT!"

The thought of the *Good Ship Lollypop* floating free in the harbor with Miss Bikini its only crew was terrifying. Fortunately, our young captain came to his senses and steered the boat into open water while a good neighbor jumped into his dinghy and retrieved the floating boat hook. Marshall and John's boat, just entering the harbor, was now in position to snatch the unclaimed buoy. However, they had witnessed the aborted attempt just described and backed off, allowing *Lollypop* a second, more controlled mooring. What good guys they were. Marshall had to use the "A" word, but he was successful in anchoring their boat for the night.

At a group gathering later that night in an on-shore restaurant, I met the young Adonis and his lovely companion. Both were from Lima, Peru, spoke excellent English, and laughed at themselves for their entertaining arrival. Both had known each other only briefly when Michael (his real name) asked Christina (her real name) if she wanted to go sailing. Love was in the air.

The next day we enjoyed some of the most extraordinary snorkeling I have ever experienced. Loblolly Bay on Anegada is surrounded by a pristine, white sand beach. You can wade sixty yards offshore in waist-deep water if you're careful not to step into the openings to underwater tunnels that lead to deep, open caverns. Snorkeling into the caverns from the surface, you can watch colorful tropical fish and coral, and still comfortably keep the entrance hole in sight. I had never seen such sights.

That night became a progressive dinner, as each boat crew laid out a selection of snacks on deck, and neighbors motored over in their dinghies for drinks, nibbles, and conversation. Great fun, although some crews, after a long night of partying, reported difficulty finding their way home in the dark.

The next day we departed on our own, some boats returning to Virgin Gorda, and others heading toward other BVI islands. After sailing southwest for an hour out of Anegada, finding our way to other islands was easy.

BVI Sunset

***

In 2005 we returned to the British Virgin Islands and had Doug and Carol, who had vacationed with us on *Sailbad the Sinner*, with us once gain. Our boat was a forty-two-foot, center-cockpit boat named *Obsession*. A center-cockpit boat has the advantage of its main stateroom aft having high headroom, since the topside cockpit is forward. It has a queen-sized bed with open space on three sides, very unusual for a typical mid-sized cruising sailboat.

Obsession's port of registry was Nice, France. Many of the Moorings boats are owned by individuals who lease them back to the charter company. I gave some consideration to this program. The plan involves buying a $300,000 boat, financing the purchase. Moorings leases the boat for four years, covering the owner's mortgage cost. They maintain the boat in top condition. During those years the owner gets six weeks each year of free sailing in any of the worldwide Moorings bases. At the end of four years, the owner either takes possession of the boat or sells it to one of the other charter companies in the BVI that purchase older boats. With luck, the selling price covers the balance of the loan. The biggest benefit to the boat owner is the free sailing during the lease-back period. *Obsession* was apparently owned by someone living in Nice, France.

Obsession, BVI, 2005

The four of us feel like old hands because of our previous visits to the BVI. Nevertheless, departure day is always a hectic period, with the boat checkout, provisioning logistics, and settling into one's boat. We shoved off as soon as possible that first day just to get into a breezy island harbor, strip down to tropical barefoot attire, and clear our heads of business and everyday routines. Exhale, relax, and then mix up a round of painkillers.

We decided Carol has a new nickname. Whenever Doug tells a story, he has to stop frequently to ask Carol, "What was that guy's name?" or "What year was that?" or "Where were we again?" I said he reminded me of the fellow who told his friend he had eaten at a fabulous restaurant and it only cost ten dollars. His friend asked the name of the restaurant. The man said, "I knew you would ask that, so I gave myself a clue. What is the flower that is generally red and has thorns on the stem?"

His friend answered, "You mean Rose."

"Yeah, that's the one. Hey, Rose, what was the name of that restaurant we ate in?"

We now call Carol "Rose."

One of the early problems we had was getting our dinghy's outboard motor to lock up in a raised position when being towed behind our sailboat. The locking pin wouldn't fall into place. I struggled with the mechanism, trying to get the motor to stay raised, when it suddenly slammed down on my hand, catching one of my fingers in its jaws. My finger wasn't broken but was sliced to the bone. Carol is a nurse, fortunately, and was able to bandage me up using the boat's first aid kit. But she said, "You need stitches."

We were on the island of Great Harbor, thinking we were miles from medical services but needed to find additional first aid supplies. A shop owner said, "Oh, we have a clinic. Go see Derek."

Derek Sweeny, a nurse practitioner from Scotland, had a two-year contract with the British government to provide medical services to the island's two hundred residents, as well as cruise ship passengers visiting the island. Asking Carol to assist, Derek gave me a tetanus shot and a dozen stitches. When I tried to pay, he said, "My services are paid by the British government."

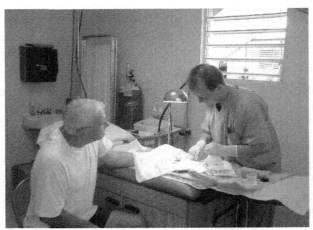

Surgery by Derek, 2005

I insisted he treat himself to a nice dinner and gave him a hundred-dollar tip. Nurse Carol was instructed to keep my wound dry for forty-eight hours and to change bandages daily. What a great crew I had.

We sailed to Virgin Gorda, stopping for an obligatory visit to the Baths. Christopher Columbus gave Virgin Gorda its name because he said it looked like a fat virgin. Go figure. Unlike previous visits, onshore waves at the Baths were eight-foot breakers, and taking the dinghy to shore looked risky. After securing *Obsession* to a mooring buoy, we opted to swim to the beach. Carol went first and Doug went last. Since Michele wasn't a strong swimmer, she and I put on life jackets and paddled in, with Michele holding tight to my vest. Close to shore the breaking waves tossed us forward, but before we could gain footing in the shallows, the backwash pulled us out into deeper water. It took four attempts before we were finally able to scramble onto the beach.

There on the beach we found Carol having her hair braided by a native. "Do you have twenty dollars?" she asked.

"No, I didn't happen to carry my money with me when swimming in," I replied.

"Where's Doug?" was her next question.

"Don't know. Haven't seen him."

We needed to bail Carol out of her purchase obligation, so I dove into the surf and was fortunately carried out beyond the breakers and able to swim back to the boat.

Doug was on board but didn't look too good. He had been tumbled repeatedly when swimming ashore and thought he was going to drown. He reached shore but then immediately retraced his steps and returned to the boat, where he was slowly recovering. No way was he going to swim back with twenty dollars for Carol. He and I took the dinghy, found a safe anchoring spot, paid Carol's ransom, and returned to *Obsession* for drinks and harrowing stories.

Since Michele and I had sailed to Anegada with the flotilla before, we felt confident we could find the island and safely enter its harbor.

On our last trip to Anegada, there were only twelve mooring buoys. The Sunsail Charter Company regularly led a flotilla to Anegada on this day, so it was important to beat them to the harbor. We won, easily beating the other Sunsail boats.

We wanted Carol and Doug to see Loblolly Bay and its incredible snorkeling site. The four of us spent two hours exploring the coral reef and its fish of many colors. After an afternoon of sun and salt water, Doug and I discovered a couple hammocks stretched between the shaded palms and caught an afternoon snooze. What a tough life this is. It just doesn't get any better than this.

After returning to the boat, then going ashore that evening, we met and visited with a few of the local residents. They all talked about Mama Anna Gada, their name for the island's nurse practitioner, like Derek on Great Island. "She is fantastic," they all said, "visiting each home daily and asking about our health. If ill," they claimed, "she would take your pain away." That night we enjoyed the largest and tastiest lobster ever imagined.

After we left Anegada the next day, we set sail for Norman Island, six hours away. No sooner had we departed than we saw a squall line ahead, with shafts of heavy rain and obviously strong winds. Four other Sunsail boats sailed close to us.

"This is going to be a blow," Doug said. "Do you want me to drop the sail?"

"No, not yet. Let's let it get a bit closer."

None of the Sunsail boats showed any sign of preparation. As the front moved closer, the winds increased our speed by several knots. "Now?" said Doug. "Shall I lower the sails now?"

"No, hold on, we have another five minutes before it hits."

Batten down the hatches

Then it was time. "Drop the sails, I'm starting the engine." With only a minute to spare, the wind hit us, heeling the boat to twenty degrees. The rain lashed the deck, obscuring all but twenty-five yards forward. Off to the side, I saw the Sunsail boats had not lowered their sails and were flailing about in the blast. Whether the crews had hesitated too long or were too fearful to go forward to lower sails, the winds whipped their boats about like leaves in the thunderstorm. The closest boat was thrown into an inadvertent jibe, heeling over to sixty degrees with sails violently lashing the deck. These were conditions that crack skulls and cause broken bones. There was nothing we could do to help but saw that the crew moved about and seemed to be okay. We continued on under diesel power.

Our destination was the Bight in Norman Island. That would put us within two hours of the Moorings base where we needed to turn in the boat at noon the following day. The Bight was also the location of the *William Thornton*, a ninety-three-foot lumber ship, turned into a floating restaurant. The *Willy T*, as it is known, is a great spot for fun and good times. It is said that if you jump naked off the deck of the *Willy T*, you get a free T-shirt. We were eager to see if anybody took that challenge, but no one did.

We also learned that the next morning a dinghy would be stationed outside the Bight with a photographer on board. If you wanted a picture of your boat, you needed to raise sails early and smile for the birdie. Photos would be posted on the internet for sixty days for anyone wishing to make a purchase.

The next day, having raised the sails, we cruised past the photographer smiling broadly.

It had been a wonderful eight days. The recipe had been simple: good winds for spirited sailing, a few squalls for excitement, newly discovered anchorages for intrigue, interesting island surgery for storytelling, and fun conversations and companionship for memories.

It just doesn't get any better than that!

# # #

Chapter Twenty-Five

Running on Empty

I ran nine marathons during my running years. Never the same one twice and they included marathons in New York, Chicago, Detroit, and Boston. All were different experiences, but one stands out because I had been told by medical professionals that if I ran, I could expect to collapse.

In 1984 I had trained for the St. Louis Marathon, but the night before the race St. Louis had a snowstorm, and knowing the streets would have difficult footing, I bowed out. However, I didn't want to waste my preparatory training, so I signed up for the Toledo Marathon, scheduled two weeks later. The Toledo Marathon started at the University of Toledo and ended at the University of Bowling Green.

The week before the event, I found myself walking through the Columbus Indiana Commons and saw that the Red Cross had set up for a blood drive. Good guy that I am, I rolled up my sleeve and donated a pint. Later that day, it occurred to me, was that a smart thing to do before running a marathon?

My next-door neighbor, Dr. Dennis Stone, had run with me on training runs over the years. He was out in his yard that night, so I walked over to tell him what I had done.

"Oh, Jim. You're not going to be able to run that race. Have you ever heard of blood doping?"

"Kinda."

"You just did the opposite. Instead of injecting oxygenated blood into your system, you gave away the red-cell-rich blood you have been building these past weeks. You're going to crash and burn."

"You've got to be kidding?"

Dennis paused just for a moment to consider giving me a blood transfusion but quickly let his medical ethics squash the idea.

"Don't run. It will be a disaster," he said.

"Well, nuts! I've trained for months, then cancelled a marathon once already. I'm going to go. If I collapse, so be it."

The Toledo Heart Stoppers Marathon held a running clinic the night before the race, presented by the faculty of the Ball State Human Performance Laboratory. After the talks, I cornered one of the presenters, told him about my blood donation and asked what he thought might happen.

The guy just turned his back and walked away.

"Hey, wait. I'd like an answer."

He stopped, turned, and said, "You're going to die."

I assumed that was an exaggeration.

I met two people that night, one a middle-aged spice collector, Frank Sprangle, who worked for Heinz. Frank traveled the world buying spices for the Heinz condiment products. Steve Chen was the other, a fifty-year-old Chinese man who had run scores of marathons, sometimes two in the same week. Steve's racing number in each race bore the number of races that event represented. The three of us chatted during the clinic and planned to have dinner together that night. Traditional pre-marathon food is pasta, for its high carbohydrate content. When I suggested an Italian restaurant to my new friends, our Asian companion said, "Ah, no. Not pasta. Must have rice! Rice is ideal food."

Well, who should know better? We ate at a Chinese restaurant.

My mania for documenting everything related to running had led me to tape-recording my marathons two years before. I Velcroed a small microcassette tape recorder to my hip and pinned a microphone to the collar of my running shirt. The original purpose of the rig was to record my split times at every mile. However, I then started interjecting the sights I saw, my physical condition, and my feelings during the race, and found the tapes to be fascinating.

Frank wanted to run with me that day, but our Chinese friend said he would do his own thing; he always brought up the rear and jogged at a very comfortable pace. I told Frank of my impending doom, and he said he'd be happy to notify officials where they might find my body.

The gun went off, and Frank and I started at a brisk sub-eight-minute-per-mile pace. Pace-wise, a ten or twelve-minute-per-mile jog is one most fit individuals find they can run comfortably. An eight-minute pace generally makes one start breathing hard. I had qualified for Boston in 1979 as a forty-year-old by running at an eight-minute-per-mile pace. A later change by the Boston rule makers, in order to limit the number of participants in the Boston race, lowered the qualifying time for over forty-year-olds to three hours and ten minutes — a seven-minute, fifteen-second pace. I hoped, with continued training, to be able to requalify one day.

Frank became fascinated with my recording of the split times. Soon, he was leaning in and saying, "Tell it Frank is doing okay, too. Little bit of a side-ache maybe." The miles were zipping by, and I was running strong and painlessly. The course had a few hills, and on those I slowed a bit, but most of my other miles were all under 7:30. Frank began tiring. This was faster than he normally ran, and at the ten-mile mark he dropped back. I still flew along, without a hint of impending doom.

The dreaded "wall" in a marathon occurs at about the twenty-mile mark, when the body's glycogen depletion is said to occur. Glucose fuels the muscles, and it's at that point some runners have an overwhelming sense of heaviness and loss of energy. I have never had that experience, but it was at the twenty-mile mark that I suspected my bloodletting crash might occur.

I clipped right along. My miles from twelve through sixteen were all under 7:30, with the sixteenth being 7:12. If I crashed at twenty, this certainly had been a great way to go.

It was getting closer. Oh, my God! I ran 7:03 on the eighteenth mile. Nineteen was slightly slower, but I held my breath looking for that next marker; and there it was, the big two-oh.

There's another phenomenon that occurs in running, called a "runner's high." It's a sense of euphoria or invincibility. Some say it's caused by the body's release of endorphins, a morphine-like chemical. Others claim it's caused by dopamine and serotonin. Whatever it is, I've never had it. I've felt good while running, but never euphoric or invincible.

The author on the run, Toledo Marathon, 1980

The twentieth mile was incredible. Never had I experienced that kind of energy. I was flying! My split at the end of twenty was 6:17! "Okay, Stark," I told myself. "Slow down. You're still six-point-two miles away." But I felt fantastic. My God, it was wonderful! Why, I thought, I'm invincible!

Those last six miles passed like a dream. I was cruising. When I listened to the tape recording later, it was unbelievable. I sounded intoxicated. I shouted to everyone I saw, "This is great! I feel wonderful! I'm flying!" My last mile was run in 7:06. It was my best marathon ever. Three hours and eighteen minutes, just eight minutes off the per-mile pace to requalify for Boston.

So what happened? What about that donated blood? I conferred with my doctor friend when I returned home, and he was astonished. Did I sneak in that transfusion after all?

Then I hit the wall big-time. I discovered on my next three-mile training run I could hardly get home. Totally exhausted, I dragged around like that for weeks. Finally, going to my regular physician, I discovered I had become anemic. It took three months to get my energy back.

And here I thought I had discovered a secret method of running your fastest marathon, running on empty.

#

Chapter Twenty-Six

Decathlon Aerobatics

Several years after my Navy days, I got a hankering once again for the thrill of aerobatics. My Cessna Skylane was not stressed for those G-forces, but there were planes around that were. I was even giving some thought to purchasing one and entering aerobatic competitions.

    First, however, I thought it might be wise to do some refresher training to sharpen my skills and see if it was all the fun I remembered. A flight instructor in Madison, Indiana, had a Super Decathlon and offered aerobatic instruction. The Decathlon, stressed for aerobatics, was a small aircraft that looked like a harmless Piper Cub with two tandem seats, one in front of the other. It was a tail dragger.

    A tail dragger has a wheel under its tail instead of a nose wheel. Airplanes of this type are tricky to land. Your airspeed at touchdown has to be precise; too slow and you'll fall out of the air, too fast and you'll bounce down the runway like a galloping horse. Landing in a crosswind takes an entirely different technique. The Super Decathlon my instructor owned had a powerful engine and was designed for stunt flying. I flew down to Madison in my plane, then practiced aerobatics in his Decathlon.

    The Experimental Aviation Association (EAA) has an International Aerobatic Club (IAC) within its organization. The EAA holds competitive events in five different categories of competition, Primary being the one I would start out in. There are different prescribed maneuvers for each category to perform, such as Cuban Eights, hammerhead stall turns, Immelmann turns, loops, spins, tail slides, and snap rolls.

I was enjoying my flights with Shawn, the instructor, and found I returned quickly to the proficiency and enjoyment I knew in the Navy. Unfortunately, because of insurance restrictions, I wasn't able to fly Shawn's airplane solo. But Eagle Creek Airport in Indianapolis had a Decathlon available to rent. You weren't supposed to do aerobatics in their plane, but after flying off solo, who would know?

The Eagle Creek Decathlon was a pretty little plane with red and white sunburst-painted wings. A Swedish Rotary exchange student, Helen, was living with my family at that time. We decided I ought to bring the Decathlon to the Columbus airport where we would pose in front of the airplane for a Christmas picture. It would also be an opportunity to take Helen and my three sons for a ride, one at a time, demonstrating loops and rolls.

According to plan, on a Sunday, while the family was in church, I flew to Eagle Creek and brought the Decathlon back to Columbus. I arrived forty-five minutes before church let out, so I tied the plane down and went into the terminal for coffee. The airport restaurant was busy that day, and I saw a number of folks gazing out the window, admiring my attractive, colorful airplane.

When the family and Helen arrived, we all posed for pictures.

(L to R) Chris, Helen, Brian, Eric, Mom, Dad

Then it was time for airplane rides. Seven-year-old Brian would be first. It's an FAA requirement that to fly aerobatics you must wear a parachute. With all those restaurant faces pressed against the glass, I cinched young Brian into a chute and showed him how to roll out of the airplane and pull the rip cord. An interesting sight to watch, I'm sure.

We climbed into our seats and strapped into our six-point shoulder harnesses. Knowing all those restaurant folks were watching our every move, I announced, "Prop clear," flipped the various switches, and started the engine. With the propeller spinning and the plane barking and spewing blue smoke across the ramp, it suddenly occurred to me, I had not untied the tie-downs from the wings. How embarrassing! How was I now going to shut down, unbuckle, emerge from the airplane to untie, and somehow pretend it was my normal way of doing things?

Fortunately, a young lineman, seeing the situation, ran to my rescue, untied the tie-downs and gave me a salute like this was our standard practice. Bless your pea-pickin' little heart, my friend!

The three boys got their airplane rides. One loop and one roll was about all the two younger stomachs could handle. But Eric, because of his previous flights with me, loved it. He wanted to try a loop on his own, and he executed it perfectly with me talking him through the procedures.

Helen was the last to go. I could tell from the worried expression on her face that she wasn't sure she was going to like being upside down in an airplane. I said, "Helen, if you would prefer, we can go up and do a little sightseeing and not bother with the parachute and aerobatics."

To my surprise, Helen wanted to wear the parachute. She was collecting pictures of her American Rotary experience, and this was a picture she just had to have.

Helen was an eighteen-year-old, rather well-endowed young lady. And she was wearing a skirt. The parachute had straps that came up between your legs and connect with the straps over your shoulders to a buckle across your chest. I let Mom make all the adjustments and connections. Same with the seat harness. It also had two straps that came up between your legs, and two around your waist, with two over your shoulders. When properly cinched in, you almost felt you were a part of the airplane.

After we'd climbed to six thousand feet, Helen wanted to know what I had done with the boys. I mentioned the maneuvers and their names. "What's a Cuban Eight?" she asked.

"Well, first you do a half loop, but when inverted on top, you roll right side up and dive into another half loop, again rolling right side up, then completing another half loop. If we were trailing smoke, we would draw the number eight in the sky."

"Ooooh, that sounds exciting. Let's do one!"

"Okay, Helen, now when we climb in the loop, tighten your stomach. Here we go!"

Diving for airspeed, I pulled the airplane nose high, and heard a continuous squeal from the back seat. After flying onto the airplane's back and then rolling right side up, I hesitated before the next dive, shouting, "Helen, are you okay?"

"Yes, yes, keep going!"

Her squeal was not one of terror but of delight, like you hear on a roller coaster. We did three Cuban Eights. Each time we did that 3-G climb in the half loop, I heard that squeal. Helen was loving it.

"What's that hammerhead thing you mentioned?" she asked.

"It starts like a half loop, except I hold the airplane straight up. It feels like the plane is coming to a stop, but just before it does, I do a half roll and let the airplane's nose fall, pointing straight down. Once we pick up speed, I pull up to level flight."

"Oooooh, yes! Let's do one."

"Helen, you're a tiger. Hold on, here we go."

Helen wanted three of those.

Helen had flown with me before when I had taken her to a Rotary meeting in Terre Haute, but I had no idea she would enjoy aerobatics like this.

"Okay, Helen, for the last stunt, *you* are going to do a loop. Grab the stick; I'll be backing you up. Just follow my instructions. Okay?"

"Oh, yes!"

"Okay, push the stick forward and pick up 180 knots. Push more ... that's it ... now pull back ... more pull ... good. Now hold it. Look out the window, Helen, make sure we are going straight up ... excellent ... here we go ... over the top. Look up, Helen, through the Plexiglas ceiling ... watch for the ground ... there it is ... make sure the wings are level ... okay, relax pressure on the stick ... down we go. Good job, Helen! Now pull back on the stick to level flight ... little more right rudder. Excellent, Helen. You did it!"

What a great day, other than my embarrassing start-up with Brian. We got our Christmas picture, the boys had their exciting airplane rides, and I was blown away by our exchange student's enjoyment of aerobatic flying.

That same Decathlon became the subject of another memorable moment that occurred on a day of frustrating matters at the office. I had been dealing with customer problems and employee issues. That afternoon I needed to get away, clear my head, and perhaps have some fun. I called Eagle Creek Airport in Indianapolis to verify availability of the Decathlon, then flew up in my plane for a session of hanging in the straps in my favorite aerobatic aircraft.

I strapped on the parachute and flew well north of metropolitan Indy so my maneuvers wouldn't be the subject of the six o'clock news. But then this incident occurred that I feared *would* make the papers.

After doing a few loops and Immelmann turns — a half loop during which you roll right side up at the top of the loop, I decided to try a couple slow rolls. Slow rolls are even more precise if doing a hesitation roll, where you pause momentarily every forty-five degrees of roll as the aircraft rotates around its horizontal axis. Done correctly, no altitude is lost or gained during the maneuver.

Although I executed these stunts above six thousand feet, I was not as sharp as normal that day and lost three hundred to four hundred feet whenever inverted. "Come on, Jim," I told myself, "you're better than this." It's not comfortable when upside down in an airplane. You experience negative G-forces, where the blood runs to your head. Perhaps, I thought, I was resisting those forces and unconsciously letting the aircraft fall when inverted.

"Okay," I said to myself, "I'll show you. I'm going to roll inverted and climb five hundred feet upside down." The Decathlon has an inverted oil system that permits the engine to keep running in that configuration. This would create some extreme negative Gs and take some strong gut-tightening to keep from passing out.

After I rolled inverted, with dust dropping from the floor above me, I pushed the stick forward to gain altitude. Whew, that really was uncomfortable; my ears and eyes felt like they were going to swell up and pop. Suddenly, from somewhere behind me, came a sound like a rifle shot. "My God, something broke! Was it a spar in one of the wings? Part of the fuselage? Did the tail break?"

With my shoulder harness cinched so tightly, I couldn't turn around to look toward the back. Relaxing my push on the stick, I let the plane fall, gently, oh so gently, bringing it right side up. All the while I expected to feel or hear the rush of wind through the torn fabric of the damaged airplane. Yet I detected nothing.

Still at six thousand feet, with the engine at idle, I floated downward in a lazy circle, looking for an open space in which to crash. And there, right under my nose, was the Zionsville airport. One single, beautiful runway, inviting me down, without having to add a bit of power or put any additional stresses on the plane.

I made the softest landing of my life. Never touched the brakes, just let the Decathlon coast to a stop on the runway. After I undid the harness and looked around, I saw nothing flapping, gaping, or hanging loose. I taxied to a parking spot, shut down, and prepared to get out for a closer inspection.

Then I saw it, on the floor by the rear seat, my clipboard. It had the paperwork I had signed when renting the plane. It had been put in a zippered pocket on the side of the cockpit. The zipper had broken. The Decathlon has a transparent Plexiglas roof so the pilot can see out when doing loops and other upside-down stunts. When the plane was inverted and pulling all those negative Gs, the clipboard shot out of the side pocket and banged against the Plexiglas ceiling with a sharp, explosive sound.

No harm done, other than soiled underwear. I returned to Eagle Creek. I had done enough flying for one day.

#

Chapter Twenty-Seven

On the Road Again

One hundred ten thousand soldiers faced each other in deadly battle on April 6, 1862. All found themselves inexperienced fighters, as this was the earliest major battle of the Civil War. Confused by poor and conflicting communications, the men fought with frenzy and violent hysteria. The fighting was ugly, said to resemble Armageddon. It lasted only two days, but casualties numbered 23,746. The fight became known as the Bloody Battle of Shiloh.

I always harbored a fascination with the Civil War. Therefore, with great eagerness I set my sights southeast from my Indiana home on a motorcycle/camping trip, planning to visit a number of the war's battlegrounds. It was 2005 and the second of such extended motorcycle trips. I was back on the road again.

Just can't wait to get out on the road again,
My apologies to Willy Nelson for borrowing his lyrics, but that song resonated in my brain from the start of my trip. I headed for the southern border of Tennessee, the first of the Civil War battlefields I intended to visit, Shiloh National Military Park. After that I would visit Stones River, and then on to Chickamauga. There would be other non–Civil War points of interest along the way, but the War between the States was my primary interest. Chickamauga held a special reason for my trip.

I had spent several years researching the history of my great-great-great-grandfather Captain Ebenezer Fletcher Stark. Eben, as he was called, had a historically significant youth when he became one of the early pioneers of an area south of

Lake Erie known as the Western Reserve. Eben arrived there from Connecticut with his parents as a sixteen-year-old in 1814. General Moses Cleaveland first surveyed the land in 1796. Stark went on to become a steamship captain on the Great Lakes. My research would eventually result in a historical novel titled *Great Lakes Skipper*. One of Eben's sons, my namesake, James Stark, enlisted in the Civil War and was killed at the Battle of Chickamauga.

I have James' military records. Included in those documents are the papers his wife, Anna Clark, filed to receive her pension for support of their children. She received eight dollars a month and two dollars for each minor child.

Each winter when Michele and I would drive south for our winter escape, we would pass through Tennessee and see signs for the Chickamauga Battlefield. Tempted to stop, we were always in a hurry and never did, but now I looked forward to trooping the grounds of my great-great-great-uncle's last stand.

First on the agenda was Shiloh. The grounds of the Shiloh Battlefield are beautiful. The rolling hills and lush forest adorned with monuments suggest a hushed reverence. Cannons cresting those hills remind you of its more clamorous past. Shiloh became the westernmost major battle of the Civil War. It was significant because early Union victories suggested the war might be a short one. However, at Shiloh the Confederate forces launched a surprise attack on Grant's army, forcing them to retreat with great losses that intimated the southern army might reverse the balance of power. However, reserve troops reinforced Grant, and he counterattacked, defeating the exhausted Confederate soldiers. The staggering losses by both sides presaged the long and brutal war that followed. The Shiloh battle is known for its Hornet's Nest, a violent skirmish in which opposing cannons fired point blank at each other; the Peach Orchard, a place where rifle fire became so intense it stripped all the peach blossoms from the trees, making the orchard appear as if in a snowstorm ; and Bloody Pond — where both wounded North and South soldiers crawled for relief and hydration, the pond soon becoming crimson from the blood.

They say that those who died at Shiloh surely went to heaven, because they had already spent their time in hell.

Bloody Shiloh

Via the beautiful Natchez Trace Parkway, I traveled east to Murfreesboro to tour the Stones River Battlefield. The fighting at Stones River has the distinction of being the deadliest battle of the Civil War. Casualties totaled about the same as those at Shiloh, but at Stones River, the 23,525 who fell represented a third of the total men fighting. Stones River in central Tennessee was a fertile belt of rich farmland, giving whichever side became victorious rich resources — crops to feed their troops. In addition, the Nashville Pike Railroad and Nashville and Chattanooga Railroad ran through Murfreesboro, offering key transportation routes to major cities in the South.

General Rosecrans of the Union army won the battle, but not decidedly so. By the end of day one of this three-day fight, the Union forces retreated. The Confederate army sensed victory, but to their surprise found the Yankees still on the battlefield with the dawning of day two. After regrouping, the Yankees drove the Rebels from the battlefield.

The "Slaughter Pen" at Stones River became famous as the place where Union forces turned back the Rebels time after time. Twenty-five percent of all the deaths at Stones River occurred in the Slaughter Pen. A small stand of trees named the "Round Forest" was defended by the Yankees and held by them throughout the entire battle. The Confederate fighters mounted attack after attack against the Round Forest, all unsuccessful. The open field before the Round Forest became known as "Hell's Half Acre."

Chickamauga National Battlefield, one of the locations Corporal James Stark fought on September 19, 1863, the day he was killed

James Stark, the brother of my great-great-grandfather Henry Stark, enlisted in the Union army in Cleveland, Ohio, in January 1863. Attached to the 124th Ohio Volunteers, the unit left immediately for duty in Kentucky and then, in February 1863, moved to Tennessee, where it was assigned to the Second Brigade, Second Division, XXI Army Corps, Army of the Cumberland. Brigadier General William B. Hazen was in command of the Second Brigade. I found it beneficial to know the details of Stark's assignment. When I arrived at the Chickamauga National Battlefield, the night before I toured the grounds, the park rangers were able to give me a print out of the troops' movements during the three days of the battle. From my records, I knew James Stark was killed on the first day of the fighting. Troop positions were marked with monuments throughout the park, and I felt certain that by standing at each of those monuments, I would feel some extrasensory connection and know the place my namesake fell. The park service also gave me a handsome Certificate of Union Service for Corporal James Stark of the 124th Infantry, Company F, Ohio.

My tour of the battlefield the next day began with the viewing of a twenty-five-minute film of the Chickamauga battle. The Union army was under the command of General Rosecrans, the same general who led them to victory at Shiloh and Stones River. He was not so fortunate at Chickamauga, however. The defeat seems to have been caused by miscommunication. At one point in the fighting, Rosecrans was told, erroneously, a gap existed in their line of defense. Believing the message to be accurate, he pulled reinforcements from other fortifications and, in fact, created a gap with their reassignment. The Confederates became surprised and elated to find a huge unprotected hole in the Federal's defenses and took full advantage of the opportunity, penetrating deep into the North's front line. The Union army never recovered.

I found the various monuments and tablets that marked Corporal Stark's movements on September 19, 1863, but I can't honestly say I became aware of any confirming vibes where he died. However, the place I think is the most likely, reads as follows:

This Brigade became engaged at 12:30 pm, northwest of the Brock Field and steadily fought its way to this line, where the action continued with severity until after 3 pm. At that hour Hazen's Brigade was relieved by Turchin's and retired to the vicinity of the Poe House to obtain ammunition.

Nearly 35,000 men made an incredible sacrifice in the woods and fields of Chickamauga. James Stark became one of the many.

Working my way east, I looked forward to reconnecting with my high school classmate Tom Catalano. Tom and I played football together at Fairfield College Preparatory School. He played guard on one side of the line, and I was guard on the other. The tackle beside Tom was Bob Malstrom. The tackle beside me was Jim Hellauer. The four of us became known as Jim and Jim on one side and Cat and Mouse on the other.

Some years back I became curious about my high school teammates and went on a search to find them. Tom was one of the early finds. That led to an enjoyable email correspondence (Tom is a great storyteller) and later to a football team reunion in West Palm Beach, Florida. Cat had a twenty-year career in the Army that included two tours in Viet Nam. He and his wife live in Elizabethtown, North Carolina, only a short distance from Fayetteville. Fort Bragg is in Fayetteville, home of the Army's Special Forces Headquarters as well as the 82nd Airborne. Fort Bragg has a museum, and who would be a better tour guide than my old friend Tom. Coincidentally, Catalano also rides a Goldwing motorcycle. We planned to meet at the museum, do a tour, and then both ride our Goldwings back to Tom's house, where I would spend the night.

The 82nd Airborne paratrooper corps had an illustrious fighting history, including landing behind enemy lines at Normandy during the D-Day invasion. Since that time the 82nd has had a part in every significant military action. Fort Bragg is also the home of the Army's Special Forces, formed by President Kennedy. Special Forces include Army Rangers, Green Berets, and other specialized unconventional warfare groups. Documentary films, dioramas, cutaways of an actual C-47, a Sherman tank, and weapons of all description made for a fascinating tour. It was especially interesting having such a knowledgeable guide telling additional stories about everything I saw.

Afterward, we rode over to Tom's lovely home for a delicious cookout, laundry services provided by his accommodating wife, an evening of engaging storytelling, and a comfortable night's rest.

The next day I rode to Wilmington, North Carolina, where I toured the battleship *North Carolina*. Wilmington acquired the ship in 1961 and has it on display. The ship has a proud history. It participated in the landing at Guadalcanal, the battle of the Philippine Sea, the landings at Iwo Jima, and the battle of Okinawa. It was an interesting visit.

I took the ferry to the Outer Banks and stopped at Kitty Hawk to tour the Wright Brothers National Monument. On December 17, 1903, the brothers made aviation history by completing the first powered aircraft flight, covering a distance of 120 feet. Much farther flights followed. What impressed me was that the brothers learned their flying skills by trial and error. Neither had any knowledge of stalling, ground effect, or the need to flair on landing. On the hundredth anniversary of the flight, an exact replica of the Wright Flyer was constructed, and skilled test pilots were invited to fly it. The best any could do was a flight of ninety feet thirty feet shorter than Orville's first flight. Amazing.

Wright Flyer, Kitty Hawk, NC

# # #

Chapter Twenty-Eight

Pamplin Historical Park

It's appropriate that I wrapped up my southeastern motorcycle trip in Petersburg, Virginia. I had started my visit of Civil War battlefields at Shiloh, one of the early engagements of the war, and Petersburg, one of the last. The breakthrough battle at Petersburg on April 2, 1865, was followed by Lee's surrender five days later.

Pamplin Historical Park, Petersburg, Virginia

Pamplin Historical Park is the site of the breakthrough battle that led to the Confederates' loss of Richmond and to Lee's capitulation. Robert Boisseau Pamplin, Jr., the former president of Georgia Pacific, and his father bought this property in the early 1990s. Their nonprofit foundation has spent $30 million developing it into what is considered America's premier Civil War history attraction. It was the most impressive sight I saw on this trip. The grounds include the National Museum of the Civil War Soldier, Tudor Hall plantation, Battlefield Center, Military Encampment, and the Breakthrough Battlefield Trail.

The museum is dedicated to the common soldier of the Civil War, both North and South, and is a most unique facility. When you enter, you are asked to choose from pictures and brief descriptions of one of thirteen soldiers to be your guide. You are then given an audio player to listen to. As you enter the various rooms and displays, the audio player automatically turns on, triggered by that display. My soldier was Lieutenant Woodcock from the Pennsylvania Volunteers, who explained his activities relating to the display before me. Sections of the museum dealt with camp life before battle, being on the march, trial by fire, death on the battlefield, surviving the field hospital, winter camp, and reenlistment.

Each section has life-size dioramas, set before walls painted with realistic murals. "Camp Life," for example, included a painting of a campground in the background, with actual tents and equipment in the foreground. Soldier mannequins huddled under a makeshift ramada were playing cards. Over the headset you could hear an explanation of early camp life with its bad food, poor accommodations, boredom, bellyaching, and training, training, training. Illnesses were another unfortunate result of living in such squalor. Outbreaks of measles, chicken pox, and dysentery infected thousands, and many died from such illnesses. I was stunned to learn that disease caused two-thirds of the Civil War deaths, rather than wounds from battle.

"On the March" detailed the equipment men carried, as well as the logistics of moving weapons, ammunition, and support supplies by wagon train. Lieutenant Woodcock also described his thoughts as he moved forward toward the enemy. Would he be injured or killed? Would he run? Could he shoot a gun to kill another man?

In the "Trial by Fire" display, the museum visitor walks through a tree-lined corridor, with the enemy visible ahead and pointing and firing rifles at you. Bullets whiz by your ears, cannonballs explode nearby, and cries of anguish and pain are heard all around. The clamor increases in intensity as you continue forward, giving you a genuine feeling of danger and apocalypse.

"A Soldier's Fate" deals with death on the battlefield. It is graphic. A Minié ball through the head would be a clean way to die. But most had limbs severed by cannonballs or bodies ripped apart by grapeshot or canister. Decomposition started quickly and there was a need to bury the dead as soon as possible. The soldier's comrades or the enemy buried the dead whenever there was a break in the fighting. If the body could be identified, it might be sent home. No such thing as dog tags existed in those days, so soldiers used to make their own, or pin their name and address to various articles of clothing for identification in case they were killed.

We have all seen the pictures of dead soldiers on Civil War battlefields. Did you ever notice that none were wearing shoes? They had been taken by others. Don't think of that as grave robbing, but rather as sharing. Materials were in such short supply that a dead soldier found with a good pair of shoes, or rifle, or haversack, or coat, was an acceptable resource. A park ranger told me that at the start of the battle, only the North had government-issued tent shelters. By the end of the war, the Confederate forces had as many tent shelters as the Union, thanks to dead Northern soldiers who shared their belongings.

"Surviving the Field Hospital" was a fascinating display. Because of the filth of the battlefield and the lack of antibiotics or sanitary care, a simple bullet wound often led to an amputation as first aid against gangrene. And no doubt it saved many lives. The video in this section of the museum shows a dramatization of the Civil War doctor removing a leg at a field hospital. It was very realistic and somewhat difficult to watch. He threw the severed leg out the window onto a pile of limbs that was reported to be two stories high at some hospitals.

Throughout the four years of the Civil War, the fighting generally stopped during the winter months. Moving troops and supplies became too difficult in cold conditions; therefore, armies went into winter camp, primitive at best, but some better than others. A reenactment of a winter camp was set up outside the museum pavilion.

The soldier's winter shelter often included a canvas roof. A few were nothing more than a tent, but some built with log walls and chinked with mud kept out the cold winds. Often, a log fireplace caked with mud as fireproofing stood adjacent to the shelter. Almost all had dirt floors. Try to imagine being half frozen in these pathetic hovels, but dreading warmer weather because you know the fighting will begin again.

Winter Camp

Tutor Hall plantation in the Pamplin Historical Park was an actual tobacco plantation built in 1812. The Boisseau family, ancestors of George Pamplin, owned the property. The plantation had a main house, a barn, several outbuildings, a working kitchen, and two slave quarters. During the war the Confederate general Sam McGowan, took over the plantation and used it as his headquarters. Decimated by the battle on

April 2, 1865, the property became unsuitable for living when the family returned after the war. They sold the property to a New York purchaser, who later sold it to Mr. Pamplin. All buildings have been restored and furnished with period furniture.

"The Battlefield Center" had a reconstruction of the earthwork, showing how to build such fortifications to protect against the enemy's assault. The earthwork was also a stage for demonstrations on cannon firing that occurred regularly throughout the day. Inside the Center is a separate museum with more artifacts, displays, and a theater. Concerned that the Historical Park might be seen as glorifying war, the foundation made a forty-minute film titled *This Terrible War*. It is graphic. It is horrifying. It is heart-wrenching. And it is excellent! It was professionally acted and dramatically presented.

"The Breakthrough Trail" a 45-minute loop of the battlefield if you walk it, tells the story of the turning point of the war that ended the rebellion. The Confederate army had heavily fortified the Petersburg area with earthworks and munitions. The Union army knew this. What they did not know was how few southern defenders manned those fortifications. When the North attacked the fortifications on Tudor Hall plantation, it had an attacking force of 14,000 soldiers. The defenders numbered only 3,500. To penetrate the lines became surprisingly easy, and the North continued its breakthrough and moved into Petersburg itself.

Lee abandoned Petersburg that night in retreat. His escape route and resupply points became complicated by Grant's rapidly pursuing army. Fighting by day and marching at night, Lee's army was exhausted and hungry. Nearly a third were killed or captured during those last few days. Finally, in the small town of Appomattox, Lee's forces became surrounded, and Lee met with Grant to discuss surrender.

The two generals met in the home of Wilmer Mclean. Grant asked only that the Confederates pledge not to take up arms against the United States. Officers were allowed to keep their side arms, and soldiers who owned their horses were permitted to take them home with them. Lee did not offer his sword, nor did Grant ask for it. The generous terms began the process of reunification.

General Grant and General Lee

The victors printed paroles for the Confederate soldiers to carry when they started their travels home. Printing presses were set-up in the Appomattox Clover Hill Tavern, and printers worked in relays to print 30,000 blank forms.

The Civil War — what a devastatingly monumental historic tragedy of human suffering and loss. And yet what incredible valor in the cause of patriotic duty.

This was just some of the fascinating history I experienced as I continued my travels with Rosey, my trusty steed.

# # #

Chapter Twenty-Nine

First and Last

From the start of my running fixation in 1974 and into its height eleven years later, my friends said I had become fanatical. Perhaps they were right. Competitively, I was a middle-of-the-pack race finisher — not particularly gifted as a speedster — but by my mid-forties, I had participated in over eighty running events and collected a fair number of trophies for my age-group finishes. My middle son, Chris, a high school track and cross-country athlete, often ran with me.

    I don't know where the idea came from; perhaps a book I read about the 1928 Transcontinental Marathon, known by its nickname *The Great Bunion Derby*, planted the seed. Organized by Charles C. Pyle, the nation's first sports agent, the '28 race began in California in March of the year and finished in New York City at the end of May. Two hundred seventy-six runners lined up for the start in Los Angeles for the $48,500 prize money. Several Olympic and European athletes ran among the starting throng. Fifty-five competitors finished the race. Its winner was Andy Payne, a part-Cherokee Indian from Oklahoma. The idea of actually traveling great distances on foot sounded intriguing and rather exciting.

Therefore, in 1987 when I suggested to Chris that we go to Florida over his spring break, he signed on enthusiastically. Until, that is, he learned I didn't intend to take a car. Instead, I planned we would fly to the west coast of Florida and run 160 miles from the Gulf of Mexico across the state to Stuart, Florida, on the Atlantic Ocean. My parents had a condominium in Stuart where we would be able to recover after the run. I proposed we run forty miles a day, taking four days to cross the state.

To run the equivalent of six back-to-back marathons caused my friends to exclaim, "What? Are you crazy?" Perhaps even more questionable was the suggestion that my seventeen-year-old son join me in this endeavor. Even as a cross-country runner, Chris had never run farther than ten miles at a time. With less than eight months to train, how could he possibly be ready?

Tom Osler, an ultra-marathoner, famous in the running world, ran more than one hundred miles at a time using a technique that we adopted. We would run two miles at a conservative pace and then walk for five minutes. The first time we tried it, we became amazed not only at comfortably covering a sixteen-mile distance, but that the walking break added only a few minutes to the overall pace.

We started training that fall, I made out a schedule that began with fifty-miles-per-week training runs and gradually built to eighty miles per week. I continued to do my training early in the morning, and Chris ran his mileage after school. Initially, we alternated six- and ten-mile weekly runs and then ran a long twenty-miler together on the weekend. Occasionally, a three-mile easy day was thrown into the mix, whenever needed for recovery. Weekly mileages gradually increased over the months leading up to our spring Florida run.

I had been training at that initial fifty-mile-per-week level for some time and was conditioned to what some referred to as the loneliness of the long-distance runner. In fact, I enjoyed the solitude. I once tried carrying a transistor radio but found it distracting; it interfered with my hypnotic thoughts as I pounded down the pre-dawn roads. I once made a comment that

I didn't think I'd ever had a good idea that wasn't born during those early-morning jogs. But what about Chris and his handling of that regimen? To run for an hour and a half after school had to be a challenge. I figured this would be the prime test of our planned adventure. My wacky idea would not be possible if Chris weren't properly conditioned and able to handle the training tedium. I didn't blame him if he could not do it; we would just do something else that spring.

Chris and Dad, 1987

Chris impressed me. He was at it every day, rain, sleet, snow, or ice. There were evenings when I saw him return from a ten-miler, his windbreaker iced over, knit hat covered in snow, and eyebrows white with frost. I marveled at his determination.

Our weekend-long runs together were enjoyable. I tried to make them interesting by selecting a variety of different routes. If it was a twenty-five-miler, we often drove one of our cars out to the destination, left it there overnight, then ran to it the next day for the drive home. Conversations took place on those three- and four-hour runs that would not have happened in a lifetime of parenting. What a bonding experience it was.

Over the months, fall and into winter, we ran in all imaginable weather conditions. A thirty-mile run would normally take four hours. One weekend we left the house in a light snow shower, heading toward a nearby town where a car had been parked the night before. The snow shower became a snowstorm — more like a blizzard — and six hours later, slipping and sliding in foot-deep snow, we reached the car, giddy with high-fives that we had conquered the elements.

Winter Challenges

Our run in Florida would be unsupported and had no accompanying vehicle to back us up. Any needed additional clothing, medical supplies, or toiletries had to be carried as we ran. We experimented with different lightweight backpacks during the training runs but finally settled on fanny packs that fastened around out waists. Even with cameras and water bottles, the packs rode comfortably on our hips with little side-to-side movement. The actual contents of the packs included: camera, film, tripod, credit cards, traveler's checks, airline tickets, aspirin, Bengay, blister first aid, extra shoestrings, water bottles, hats, suntan lotion, Vaseline, bandanas, toothbrush, toothpaste, comb, ChapStick, eyeglasses, case, dental floss, gum, maps, Band-Aids, pocket knife, T-shirt, extra socks, and nylon wind pants and jacket.

The two water bottles we carried held a total of forty-eight ounces of water and collapsed as the water was consumed.

The nylon jacket and pants were the outfits worn on the airplane to Florida and what we changed into at the end of each day.

The route we ran was unpopulated, with long stretches of nothing but farmland and orange groves. Fortunately, there were three small crossroad towns, conveniently located about forty miles apart along the route. Each had motels available for our end-of-the-day accommodations and rest.

Our last week of heavy training, two weeks before leaving for Florida, had us running eighty miles, with a forty-mile run on the weekend. That's the only time we actually ran that far in one day, and we found it extremely difficult. The last six miles of that forty-mile run were depressing. Not only painful physically, but the thought of attempting that distance for four consecutive days was daunting. Were we ready? After an easy week to rest up and recover, we were about to find out.

My employees at the office held a sendoff party for the two of us. They gave us gag gifts, like a compass to keep us headed in the right direction, and a large fake thumb to use to beg for a ride if needed. Someone at the party asked, "Are you ready?" And I said I sensed that we were. However, I added, "We're hoping for sixty-degree temperatures. If it's seventy, it will be much more difficult. And if it's eighty, well, that might spell trouble."

We flew into Sarasota and were picked up at the airport by friends who put us up the night before we began the run. They fixed a typical pre-marathon pasta dinner that night, then a huge stack of pancakes the next morning. Our friends and my brother-in-law, who was vacationing with his family in Florida, all gathered at the beach for the start of our run the next morning.

Someone had alerted the local newspaper, the *Pelican Press*, for the start of our event, so we had the fun of a photo shoot as we ceremoniously dipped our right feet in the Gulf of Mexico and headed off toward the Atlantic Ocean 164 miles away.

Dipping right foot into Gulf of Mexico

After our previous easy week of low mileage, I commented to Chris, "Isn't it great to be running again?" As soon as I said that, I hoped those words would not be regretted. The first two and half hours were pleasant. The warm early-morning sun felt good, and there was a lot of brave, enthusiastic conversation. Our only concern when we started was a sign that said "Arcadia, 48 miles." That was our first day's destination, and it was six miles farther than we had anticipated.

After three and half hours, the excitement of our start had worn off. The opponents were no longer distance and fitness; the enemy was the temperature. Nearing ninety degrees, the heat washed over our bodies in waves from the blazing sun. The wide-open spaces provided no shade. We had lathered ourselves with sunscreen, but my biggest concern was dehydration. Our water supply dwindled rapidly, and we didn't know where we would find more.

On the Road

Chris' labored breathing and lack of conversation told me he was particularly suffering from the heat. After four and a half hours with the sun still directly overhead, we decided to crawl under some shrubbery and wait until later in the day when the sun sank lower in the sky. We had covered only half the distance to Arcadia.

After lying low for three hours and nursing our water supply, Chris thought he had recovered somewhat and was able to continue. We modified our technique of running two miles and walking for five minutes to running for ten minutes and walking five.

After another hour, with water now gone, we came upon a plantation. The lane leading up to the main building was three hundred yards long, and although the gate was locked, I said, "Come on, Chris, we've got to find some water."

The plantation home was right out of *Gone With the Wind*, with tall white columns supporting both an upper and a lower porch. Interestingly, each porch was lined with white, cane-backed rocking chairs. After knocking on the door and getting no response, we noticed a brass plaque by the door that indicated the property was owned by the Amax Corporation. I recalled Amax was the parent company of Stanley Tools and several other brands. We looked around the property and saw a picnic area, suggesting the company used this plantation for entertaining purposes. In back of the house we found a party wagon, which, although it had no beer keg, did have a half-filled water cooler. Better than beer! After filling our empty water containers, Chris and I relaxed by quenching our thirst while rocking on the front veranda of our southern plantation.

As pleasant as that interlude happened to be, we still had several miles to go on the first leg of our travels. It was nearing 7:00 p.m., and we hoped to find Acadia just around the next corner. We approached a green sign that we prayed announced the entrance to the town. You can't imagine the depression when we read, "Acadia – 7 Miles." It might as well have said 70!

With the sun setting, Chris was done running. We walked those last seven miles, a walk that lasted two and a half hours. We dragged into the lobby of our motel, some forty-eight miles and thirteen hours after my comment "Isn't it great to be running again?"

I still recall the bizarre scene in that lobby. The two of us were filthy from waist to foot with sweat and accumulated highway dust. The clerk, without noticing, asked for our car license number. "We're on foot," I replied, and with nary a reaction the clerk said, "Okay, then. That'll be $35.00."

What? You have a number of tourists crossing Florida on foot these days?

After showering, we ordered pizza to be delivered as we prepared for bed. Troubling thoughts swirled through my mind. Was there any way we might continue this insanity? The forecast for the next day was again in the nineties. We had worked so hard to prepare for this adventure — was it over? Was there any way Chris might recover by morning? Muscles don't bounce back quickly after that kind of abuse. Even if it were possible to slug it out another day in the heat, would we be able to do it for three more days?

After two pieces of pizza, we could eat no more and collapsed unconscious into bed.

Day two dawned with the most ghastly weather you could imagine; brilliant sun in a cloudless blue sky. What we would have given for a rainy, overcast day! Though calf and thigh muscles felt like hamburger, we decided to start out to see what would happen.

We ran for two hours. How Chris could do that after the ordeal of his first day was beyond me. However, his ragged breathing proved evidence enough that he'd never last six hours that day. We crawled into the shade of a palmetto plant to discuss the options. We decided that if it were possible to get Chris a ride to Lake Placid, our next stop, perhaps he might rest enough to continue the following day. I was feeling okay so far and would continue to run, meeting him later at the motel. Chris' disappointment in not running every mile was distressing, but we both realized he had no choice at that point.

We flagged down a county highway truck that gave us a ride to a Florida prison, of all places, just half a mile from where we had stopped. I used a phone there and called a cab in Acadia to come out and take Chris to Lake Placid, where we had motel reservations. He took my nonessentials, and I took his water bottles and sunscreen, and we separated.

The first three hours passed tolerably well, but the last two and a half hours were brutal. It ended at the intersection of Routes 70 and 27 at 4:30 p.m. The motel was another mile and a half north, which I originally thought we might either walk or run to. Both of those options were now out of the question; I needed a ride.

There was a gas station at the intersection, and I just draped myself over a gas pump and started begging for a ride. I was covered in black dirt from the knees down, with two days' growth of beard on my face, algae and swamp grass hanging from my hat and shirt after dipping them in anything wet to keep cool, and I thought, Folks sure are unfriendly in Florida.

Finally, I convinced a young man who had just put two dollars' worth of gas in his car to accept my five dollars to take me one and a half miles up the road. When he dropped me off, he suggested I shouldn't run so far from the motel if I couldn't get back. I just nodded.

Chris had gotten in about noon, soaked in a hot tub of water, and slept most of the afternoon in the air-conditioned room. We decided we would try again the next day but would get up at 4:00 a.m. in an attempt to beat the heat.

After calling home to assure Mom we were having the time of our lives in the Sunshine State, we had an early dinner and hit the rack to end a disappointing second day.

Four a.m. rolled around quickly. The stiffness and sore muscles didn't seem any more pronounced than expected, so with our newly purchased flashlights, we started off in the predawn darkness. Not much conversation took place the first hour, as I listened for clues in Chris' breathing that would tell me if this trip was a bust, or if he had actually recovered since the first day. I was not able to tell.

Into the second hour I began to suspect Chris was firing on all cylinders once again. He moved smoothly at a good pace, and, most revealing of all, the conversation had taken on a positive tone. We talked about the swim we were going to enjoy in the motel's pool in Okeechobee, our next destination.

This would be our shortest day. Lake Placid to Okeechobee was only thirty-six miles. The map showed the small town of Brighton halfway between, which suggested the possibility of soft drinks or maybe the supreme joy of an ice cream cone. You get rather hungry when burning an extra five thousand calories a day.

The sunrise that morning was spectacular, a world turned golden-orange, all for the benefit of two Hoosiers a long way from home. Of brief concern was a broken strap on Chris' fanny pack, which I repaired with the extra shoestrings I carried. We hoped to find someone in the town up ahead, a cobbler or tailor, who might sew it together for us.

We arrived in Brighton at 8:30 a.m. The whole town came out to meet us; that's one mangy mutt and two Florida cowboys in a 1948 Studebaker. Brighton had one building, a grain elevator — no Coke machine, no ice cream parlor, and I didn't even ask about the tailor shop.

Brighton did have a water faucet, however, so the two of us parked ourselves in the shade of the elevator, drank water, and nibbled on granola bars. It was there that we laughed about all the fun people said we might have running across Florida together. Actually, Brighton was a positive moment, because we realized it was only 8:30 a.m. and we were already halfway to our day's destination.

Things got even better. As we left Brighton, we saw a mileage sign up ahead. We had become leery of mileage signs, because they always seemed to announce more miles than we expected. As we approached this sign, we guessed Okeechobee to be eighteen miles at worst and fifteen miles at best. The sign said Okeechobee – 13 miles! That's like Christmas and birthdays all rolled into one. We couldn't have been in better spirits for the next two hours after seeing that sign. The sign was wrong, incidentally; it really was eighteen miles to our motel, but the last five were through shaded streets, and they went quickly.

My dad drove over from Stuart that evening, planning to follow us the next day and provide support. We would run without our packs, and having his car waiting with iced-down drinks every half hour was wonderful. Also, sharing the adventure among three generations added to the event.

This route also included a twelve-mile dirt road devoid of truck traffic. It had been the pits having enormous vehicles roaring past us for three days, practically blowing us off the highway with their whirlwinds of grimy backwash.

Another plus for the day was removal of the uncertainty factor. During the previous days, we were never quite certain where the day's finish line was located. Was it just around the next bend or still miles away? After six hours of running while thinking the next town must be just ahead, emerging from a grove of trees only to see a ribbon of highway stretched out before you and disappearing in the distance, was enough to make you weep. That would not be the case on the fourth day. My dad had recorded every mile on his drive over from Stuart and knew exactly how far it was to his condo. It would be our second-longest day, but it proved a lot easier because of the above factors.

My dad appeared to really enjoy being part of our venture and spent the time between our thirty-minute arrivals recording our split times, calculating the ratio of cars going east versus those going west, reading three *Wall Street Journal*s, and pondering the question "How can two people drink two and a half quarts of water, forty-eight ounces of Gatorade, and thirty-two ounces of orange juice, and not spend all day looking for a bathroom?"

Fourth and final day

The final four miles into Stuart ran through a picturesque golf complex, over the St. Luce River, and up to the entrance of my folks' condo community. There we passed under a congratulatory banner surrounded by family and neighbors and one rather confused mailman.

My two sisters had flown down from their homes in Connecticut and Chicago to be part of the festivities and had erected the banner announcing our arrival. The condo wasn't the actual finish line, as we would finish the run at an Atlantic Ocean beach. Still, seven miles farther away, we had already decided we would wrap up the marathon the next morning with the final seven miles.

The local Stuart newspaper had written an article about our run, and a number of my parents' friends called and wanted to be at the beach when we finished. In addition, a West Palm Beach newspaper had seen the article, and they were sending a photographer over to photograph the finish. Chris and I planned our next morning's run to arrive at the beach at 9:30 a.m.

The next morning, just a few minutes into our run, we spotted a battered Volkswagen with a large telephoto lens sticking out of its side window. Chris and I sucked in our guts and gave him our best *Chariots of Fire* running form. After we passed, the photographer zoomed up ahead to position himself for another series of action shots. This time he sprawled across the sidewalk and shot our approach head-on.

Panting from our pace, I stopped when we reached this guy and explained that we were going to have to slow down or we would beat our friends to the beach by half an hour. He understood and said he had all he needed and would meet us there later on.

Throughout the preceding four days, and in fact during the previous several months leading up to our run, I had been concerned about physical injuries that might put an end to our quest — a pulled muscle, a twisted ankle, or a stress fracture. Therefore, my heart stopped when just three miles from the finish, Chris shouted, "Oh no!"

"Chris, what is it?"

"I can't believe it," he said. "My jock just broke. It's come completely apart."

We scrambled under a bridge to assess the problem, and Chris took off his supporter ... over his head.

"What's the big problem?" I asked. "Don't your running shorts have a brief sewn into them?"

"Not this pair," he said. "And with that photographer coming to the beach, it might look strange with me tip-toeing across the sand."

The solution was simple. All my clothing was fully operational. My running shorts had a sewn-in brief, plus I had an intact supporter. I would just trade shorts with Chris.

Left foot into the Atlantic Ocean

The pictures that appeared in the West Palm Beach newspaper the following day covered the front page of the sporting section. A really observant reader might notice that in the picture of Chris and me running along the sidewalk, I have the shorts with the white stripe along the leg. However, in the picture taken at the beach, after ceremoniously dipping our left feet in the Atlantic, Chris has the shorts with the white stripe.

The headline accompanying those photos read, "First and Last 164-Mile Marathon," and to that I say, "Amen, brother!"

First (and last). Amen, Brother!

#

Chapter Thirty

Hang Gliders: So Safe They Can Just Barely Kill You

The tow plane tugged at my hang glider, pulling it higher and higher toward the two-thousand-foot release altitude. Wind screamed in my ears, adding to the excitement as objects on the ground grew smaller and smaller. The briefing before the flight warned me to expect several seconds of weightlessness when cutting loose, much like the free-fall when cresting the first hill of a roller coaster. It proves to be an apt description. I gasped for breath when the glider dropped downward, seeking the lift that as if by magic will give my sailcloth wings birdlike properties

I connected myself to the fifty-pound structure of aluminum and polyester fabric with a suspension harness that hung from the glider and a horizontal control bar held in my vicelike grip. A push on that bar would raise the glider's nose, slowing it for landing. A pull on the bar puts the glider into a dive and accelerated its speed. Swinging the pilot's weight either left or right in the straps achieves the right and left banking turns. A mock trainer on the ground provided ample opportunity to practice the side-to-side weight shift. How difficult can this be? I mused. As a former Navy pilot with thousands of flight hours in complex aircraft, it seemed laughable that controlling this simple people-carrying kite would be anything but bicycle-riding easy.

When I recovered from the tow plane's release and after I established a steady flying speed, I surveyed the terrain below to consider the direction I needed to go to set up the approach for the landing. Previous practice landings when gliding off the training hill gave me confidence the final touchdown could be executed with safety when lined up with the airfield.

Deciding I must go left toward the north end of the flying area, I swung my body with confidence left in the harness. To my astonishment and alarm, I abruptly turned right! *What the hell! What's going on?* I recentered myself to stop the turn and reestablished straight and level flight. Confused by the reverse reaction to my weight shift, I tried again, this time making a slight right-side weight shift, but again with the opposite result: a left bank. I had a serious problem. The contradictory response to these intuitive and logical actions could be disastrous.

If the Navy taught me anything, it is not to panic, calm yourself, and evaluate the situation. Thinking through the weight-shift action, I discovered that although my head and shoulders swung in the intended direction, my butt and legs, the heavier end of me, tended toward the other side. A cautious test, ensuring that my entire body pulled left, resulted in a satisfactory left bank. *Whew.*

With more-focused control movements, I now circled the field, aligned myself with the landing area, and executed an acceptable landing.

Several months earlier in 2005, when planning a three-week motorcycle trip throughout the Southeast to visit Civil War battlefields, I saw I would pass near Lookout Mountain, Georgia, home of the Lookout Mountain Hang Gliding School. I had one previous involvement with hang gliders a few years earlier when vacationing in the Outer Banks. The experience involved running down a sand dune and gliding a hundred yards or so – if done as instructed. Several landings ended with a snoot-full of sand, but I delighted in the fun, and it piqued my interest in seagull-like pursuits.

It's a surprise to many that hang gliders, although powerless, can stay aloft for extended periods if near uprising thermals, such as you find in the foothills of Arizona. I flew a sailplane there one time. Some years after that, I watched hang gliders float for miles along the beaches of California on the onshore breezes. Such silent flying excited my adventuring sprit, and I enrolled for a week's course at the Lookout Mountain School.

Headquarters for the school is perched on top of Lookout Mountain. When I arrived, glider pilots could be seen stepping off a concrete launching platform and floating down to the landing field 1,500 feet below. Could I do that? I wondered. Hell yes. Look at all the pilot training and experience I've had. However, if the truth be known, without hang glider training, stepping off that concrete slab would no doubt have resulted in a tight spiral into the trees and rocks below. Thank heavens no one double-dog-dared me to try.

Lookout Mountain Hang Gliding School, Lookout Mountain, Georgia, 2005

My five-day school involved initial training on a low hill, where we practiced running off its crest into the launch and gliding to a full stall landing. Because of the beastly hot temperatures in Georgia that June, we started classes right after sunup to avoid the heat-generated turbulence. My instructor, Rex, began the first day's session by having me sign multiple sheets of release forms that seemed to suggest that hang gliding gives you a one-in-ten chance of survival. "YES, I understand hang gliding is a dangerous sport. YES, I know the FAA does not certify hang gliders. YES, I agree, I will not sue Lookout Mountain Hang Gliding School if I'm killed."

Rex was a thirty-year-old free spirit, who came to the school periodically to instruct and "do some flyin', man." I didn't find out what he did during the rest of his life. I imagined he, like numerous other young people I see in my university town, never quite settled down after college. He probably works part-time at the YMCA, or lifeguards at the country club pool, or hangs out in the city park talking about the advanced degrees he thinks about taking. I admired Rex's cool, don't-worry-be-happy demeanor. His father lived in Indiana, as I did, and rode a motorcycle. So how do you think Rex and I got along?

As a sixty-seven-year-old, white-haired flyer among my group of daredevil trainees, I stood out among my twenty-year-old companions. Will he break his neck? I imagined they all wondered.

The low hill we started on measured only two hundred yards from top to bottom. The launch procedure involved first squatting beneath the frame of the glider, attaching the suspension straps to the pilot's harness, then standing, lifting the fifty-pound glider to the shoulders. Easy at first, by the end of five days, lifting those fifty pounds made my thighs feel as if I had run five marathons. Facing into the breeze, the pilot starts running down the slope of the hill. It takes only a few steps before the glider lifts off the shoulders and pulls you forward. There's a temptation to quit running. That would be a mistake, because if the glider settles, the flyer finds himself making a painful belly flop and being dragged a hundred yards down the hill. "Keep running," the instructor yells. Pilots floating ten feet above the ground with feet pedaling madly in midair is a humorous sight to see.

Within seconds the glider reaches the bottom of the hill, and the pilot must be ready for the landing. The goal is to push the horizontal bar straight forward, bringing the glider to a full stall stop and to finish standing up. I take two days to achieve my first stand-up landing. The push never came soon enough, and the glider didn't become sufficiently slow to enable me to keep my feet under me. Because of the heat, we all wore shorts. The flopped-down, belly landings soon removed all the skin from both knees and elbows. I commented to Rex after the second day that knee pads would be a good accessory for novice pilots. "Oh, we have those," he said. "They're in the barn where you got your harness and helmet."

"Thanks, Rex. Now you tell me."

Students carry their own glider back up the hill following each flight, but because I undoubtedly remind Rex of his dad, he always refused my halfhearted protest and carried my glider for me. Some young boys appeared on the field the second day, offering their toting services to other flyers, earning tips for their help. Even without carrying the glider, the hike back up the hill in the ninety-degree heat made for an exhausting day. I was camping in the school's campground. After each morning's session, I found myself barely able to make it back to my tent before collapsing for a two-hour recovery snooze.

By the third day, I made stand-up landings with some regularity. We then moved over to a higher hill, which gave us longer flights and opportunities to practice in-flight maneuvers. The gliders we flew off the hills differed slightly from the towed models. When flying off the hills our legs hung vertically below the glider. When towed aloft by the tow plane, an installed sling supported the legs, putting one's body in a more horizontal position. That different configuration contributed to my problem with cross-controlling the weight shift when first being towed aloft. The towed gliders also had small wheels installed below the horizontal bar that kept your horizontal body above the ground at the start of the tow and when landing.

My five-day syllabus included two additional towed-to-altitude flights. Rex observed my flights from the ground and announced after each flight, "You're awesome, man." I'm not sure about awesome, but I felt more confident and relaxed with each flight. Some of my younger fellow students continued to have problems and never got off the low hill, nor did they get approved for a solo towed flight. *Let's hear it for the old guy.*

A picture of the famed U.S. Navy "Blue Angel" flight demonstration team hangs in the headquarters building. Those guys are the best of the best when it comes to flying. The story is, they came to the area for an airshow and visited the school to check out hang gliders. The manager in the office said they went aloft in two-person hang gliders but complained that without altimeters, airspeed indicators, and ejection seats, hang gliders were just too dangerous. Aw, shucks, guys. Where's your Wright Brothers sense of adventure?

I enjoyed the week but will admit I'm still not ready to step off that concrete ramp atop Lookout Mountain.

# # #

Chapter Thirty-One

Ike's Place

Mamie and Dwight Eisenhower had just settled down to watch *Bonanza*, Ike's favorite TV show, when the black, rotary-dial telephone on the side table rang. "Now, who the hell can that be?" Ike wanted to know. "You end a war in Europe, serve as president of Columbia, spend two terms as president of the U.S. promoting Atoms for Peace, establish the states of Alaska and Hawaii, and pass the '57 Civil Rights Act, and the bastards won't let you retire in peace."

"Now, Ike, it might be one of the children. I'll get it. - Oh, hello, Mr. Kennedy. – No, I'm sorry, Ike's busy, out in the barn fooling with that old golf cart of his. Can I have him call you when he gets in? – About the Cuban thing? – Okay, I'll tell him. – Bye."

"Damn, that young fella has called me a dozen times. Wants to know how to put an invasion force together. Should never make a Navy man commander in chief. I'll call him after the show."

I imagined the above scene as I stood in the enclosed side porch of President Eisenhower's Gettysburg retirement home. The guide told us how Ike loved *Bonanza* and used to watch it on their blond-oak-finish, twenty-four-inch, black-and-white Philco TV. Color sets were available at the time, but Ike just paid good money for that one, the "Director" model, with electrostatic remote control.

The Eisenhowers bought the home adjacent to the Gettysburg National Battleground in the 1950s but didn't retire there full-time until 1961. The farm is beautiful. Besides the large main house, there are several attractive out-buildings. Ike used to motor around the grounds in an old three-wheeled golf cart with tiller steering.

I was with former high school classmate Jim Hellauer, of the previously mentioned duo of Jim and Jim. I motorcycled from my home in Indiana to one of the Gettysburg campgrounds and had Jim join me for a battlefield tour, since he lived close by, near Philadelphia.

Jim and I had first reconnected forty years after our graduation from Fairfield Prep. He found my contact information in a just-published school directory and called my office. "Hey, Starky. It's Hellauer. What have you been up to for the last four decades?" Jim had gone to the Naval Academy, and we'd lost track of each other. That phone call started an annual visit: one year to his winter home in West Palm Beach, and the next year to our place in The Villages.

Jim Hellauer & Stark at Gettysburg, 2007

I'd toured other battlefields on my motorcycle/camping trip two years before, in 2005, and was eager to visit the most famous Civil War site of them all, Gettysburg. Jim met me at the visitor's center in the morning, and we had a tour guide drive Jim's car as he narrated the infamous events of July 1863.

It was a three-day battle, July 1 through 3, 1863. Confederate forces had recently claimed a major victory at Chancellorsville and felt that by attacking the Union forces up north in Pennsylvania, they might end the war with a final victory. They almost did. Gettysburg had the greatest number of casualties of any battle of the war.

The Confederates outnumbered the Union forces, but the first federal commander on the scene at Gettysburg, General Reynolds, knew if he could position his troops on high ground, he would have a decided advantage. The high ground was Seminary Ridge, so named because of the Lutheran seminary located there. Reynolds pledged to hold the ground at all costs and, true to his pledge, did, but it cost him his life.

When day two dawned, the Union forces, in control of the high ground, had a decidedly superior position. One of the other high points manned by the Yankees was Little Round Top. The commander there, General Daniel Sickles, abandoned Little Round Top in favor of a different fortification, which was later regarded a classic blunder. As the battle unfolded, both sides discovered Little Round Top was up for grabs. It became a footrace to see which side would take possession. A footrace that was encumbered by hauling heavy artillery without the benefit of horse power. The Federal forces won the race by a mere four minutes and took command of that important fortification.

On day three, the Confederate strategists thought they saw a weakness in the Union line. General Pickett led an attack that became known as Pickett's Charge. The Rebel intelligence was wrong, there was not a weakness in the Union line, and a bitter battle was about to ensue. The Federal forces, seeing the charge taking shape, ceased firing and ducked down out of sight. Perhaps the Confederate soldiers would think the Feds had been broken and would march forward into the open unprotected. It worked. Thirteen thousand southern troops marched out of the woods in a line half a mile wide.

An official report follows:
Red flags wave, their horsemen gallop up and down the ranks, the arms of 13,000 men, barrels and bayonets gleam in the sun, a forest of flashing steel. On they move, as with one soul, in perfect order without impediment of ditch, wall, or stream, over ridge and slope, through orchard, meadow, and cornfield.
It was, a northern officer remembered, the most beautiful thing he ever saw.

Union guns on Seminary Ridge and Little Round Top opened fire on the right flank of the advancing Confederate line. "We could not help hitting them with every shot," an officer recalled. "As many as ten men at a time were destroyed by a single bursting shell."

The Confederates still "came on in magnificent order," an admiring Union private wrote, "with the step of men who believed themselves invincible ... solid shots plowed huge lanes in their close columns, but their shattered lines did not waver. With banners waving, they swept on with steady step like an irresistible wave of fate."

Directly ahead, behind a stone wall, the Union men continued to hold their fire. At last, General Alexander Hays gave the order, "Open fire!" The Rebel lines were at once enveloped in a dense cloud of dust. A Federal officer said, "Arms, heads, blankets, guns, and knapsacks were tossed into the clear air. A moan went up from the field."

The fighting was as furious as any seen during the war. "Seconds are centuries, minutes ages, men fire into each other's faces," a survivor wrote, "not five feet apart. There are bayonet thrusts, saber strokes, pistol shots ... men going down on hands and knees, spinning around like tops, throwing out their arms, gulping blood, falling; legless, armless, headless. There are ghastly heaps of dead men."

Confederate soldiers of the front rank look around for their supports. They are gone — fleeing over the field; broken, shattered, thrown into confusion by the remorseless fire. Thousands of rebels throw down their arms and give themselves up as prisoners.

The tour guide driving Jim's car actually choked up describing the battle. Americans killing Americans. May our country never see such a tragedy again.

After I left Gettysburg, I continued my solo travels through Connecticut and Maine, then took the ferry to Nova Scotia, where Michele flew up and joined me for a week of sightseeing. Those days were full of fascinating sights and discoveries described elsewhere in this memoir.

My homeward leg included camping in Maine's Acadia National Park, where I tested my tent and lightweight sleeping bag against Maine's fall night's temperatures. The campground I stayed in had one hundred tent sites. "Sorry," I said, "I don't have a reservation."

"No problem," said the ranger. "You're the only tent here. It's always that way when the nights drop below freezing."

And it was no problem. My camping-expert youngest son had said to wear long underwear, a knit hat, and socks at night, and you'll do fine. He was right, but it sure made those midnight trips to the restroom hurried.

After leaving Maine, I rode into Gorham, New Hampshire, and the White Mountains. The clear blue sky was a stunning backdrop for the sight of Mount Washington reaching up over a mile high. Michele and I had toured New Hampshire on motorcycle in 1992 but didn't ride up Mt. Washington, because clouds shrouded the mountain at the time. This now was my opportunity.

The road up is steep, which is not a problem, even with a heavily loaded motorcycle, as long as you can maintain a stabilizing speed of twenty miles per hour or more. For a quarter of the way up, I had the narrow road to myself, but then I caught up with a slow-moving vehicle.

Note Chains holding building down

The hard-packed, graded, dirt road surface was safe enough, except at every hairpin curve the car in front of me slowed to five miles per hour. Coming to a complete stop on such a steep surface could be difficult, so I tried to open up as much distance as possible between me and my leader and prayed he would keep moving. At last, I made it to the top.

The summit is above the timberline and has a moonscape appearance. Several weather observation buildings are located there, one of which — an older building — had chains draped over its roofline to hold it down. The highest wind ever recorded was on top of Mt. Washington. It blew 231 mph in 1934.

I called Michele on my cell phone. "You'll never guess where I am at the moment!" And she couldn't.

After walking around and snapping a few pictures, I departed. The ride down was much easier. I put the motorcycle in second gear and rarely had to hit the brakes the entire way. The Mt. Washington road was exhilarating, but it surprised me how challenging it was. If I had made that climb fifteen years ago with Michele on the back, it would have been an exciting adventure. I doubt she would have opened her eyes during the entire ride.

A fellow motorcyclist I met on the ferry to Nova Scotia had been on a group ride and told me about a spectacular section they had ridden through in New Hampshire, the Kancamagus Pass. On his recommendation, I put it on my list of places to see. It's only fifty-six miles long, but, wanting to give it plenty of time for appreciation, plus an early-morning sun for pictures, I spent the night before in a motel.

It was brilliantly clear the next day. Even though the pass is less than sixty miles long, had it been any longer, it would have taken the entire day, because I was stopping to take pictures at every turn. The highway had sweeping turns between the surrounding mountains and followed a rushing mountain stream. Early-morning sun illuminated the surroundings with postcard vividness. The creek reflected the colors of the sky and the reds and oranges of the fall leaves. The boulder-strewn creek bed was encircled by the white-birch-flecked woods, creating breathtaking images.

That's it. Kancamagus Pass ranked as the most beautiful sight of the trip. But then, I said that when visiting those fishing villages in Nova Scotia. Then again, the coastline around Acadia National Park that was spectacular. And I just loved the harbor in Camden, Maine. And the Dartmouth College campus was classic. Oh, heck. All have their own unique beauty.

Another three-week trip with but two days of rain, fascinating sights, interesting people, and nary a mishap. The Stark good luck charm continues its magic.

#

Chapter Thirty-Two

Glacier National Park

My motorcycle rolled into the harbor community of Bayfield, Wisconsin. The masts of sailboats could be seen a block away, swaying back and forth like cornstalks in a farmer's field. Normally, harbors and boats are my favorite scenes, but this day was gray, rainy, and raw. My first thought was to find a motel and unload my hundred pounds of camping gear before I explored further.

It was 2013, and I'd been gone only a few days, on what would turn out to be a three-week trip. My westernmost destination would be Glacier National Park, but getting there would include a number of stops. Bayfield is the primary stepping-off place for the Apostle Islands, twenty-two islands a short distance from the Wisconsin shore out in Lake Superior.

A name like Apostle Islands might lead you to expect each to have the name of a different biblical disciples, but not so. Different lumber and quarry companies owned the islands over the years, and names were changed at the whim of the current owners. Today those titles are Eagle, York, Bell, and other ordinary-sounding labels. All but one of the twenty-two islands are now part of the National Park Service. Madeline Island, named after an Indian princess who married a French explorer in a Christian ceremony, remains in private hands.

Twenty thousand years before, the islands had been part of the Wisconsin landmass, but during the Ice Age and Glacier period, they became separated by erosion. First discovered by French explorers in the 1700s, they were named the Apostles by the Catholic French, as it was their custom to use holy names.

The Apostles had been on my list of places to visit ever since my sons and I first came to the state as part of our annual air show attendance at Oshkosh. We always traveled around the state after the show, camping in different state parks, but never made it to the islands.

Cruise boat, Apostle Islands, 2013

I seized the opportunity. A visitor-center person suggested I sign up for the Apostle Island Discovery Cruise. The large tour boat carried two hundred tourists. The captain did an excellent job of narrating the trip, and I learned that Lake Superior is the largest of the Great Lakes, covering an area about the size of Indiana. It is fed by 331 rivers and streams and is over thousand feet deep in places.

During our cruise we saw bald eagles in shoreline treetops and black bears frolicking along the water's edge. Bears are good swimmers and move from one island to another. The captain explained that the black bears live on berries and vegetation and are not by nature a threat to humans. However, some campers, aware of the nonthreatening nature of the bears, made the mistake of feeding the bears, some of whom acquired a taste for camp fare like hamburgers and hot dogs. The captain then told the story of one bear, known as Star because of the white marking on its face, who became so attracted to food off the grill, it started raiding campsites and helping itself. Star became bothersome, and park rangers had to capture Star and move it to an uninhabited island three hundred miles away. To everyone's surprise, early the next season, Star showed up, back at its original island, looking for barbecue. That's one Olympic swimming bear!

The Caves, Devil's Island, 2013

The caves on Devil's Island are impressive. Lake Superior is often a violent, stormy sea. Winds of 50 to 80 mph are not uncommon, creating waves twenty or thirty feet high. Devil's Island receives a direct blast from those storms. The sandstone cliffs of the island have been eroded by the battering winds and waves for hundreds of years. Water undercut the stone and created a honeycomb of spectacular caves.

Indians named it Devil's Island because winds of twenty-five mph caused the caves to emit a mournful moan. Winds of fifty mph made the caves shriek in a woman's high voice. When thirty-foot waves pounded against the caves, they would compress the air inside, and it would explode like a bomb.

The captain said he had never seen the lake as calm as it was the night of our tour and got our boat within a few feet of the cave openings for close-up photos.

The Apostle Islands are intriguing. It would be fun to return and rent a kayak for further exploration and discovery.

After crossing Minnesota, I set my sights on the border of North Dakota and the Theodore Roosevelt National Park. The picturesque bluffs and buttes found throughout the park were first formed 60 million years ago. Sediment washed down from the young Rocky Mountains covered the prairie. It consisted mostly of soft sand and clay, and when seams of coal ignited from lightning strikes, it baked the material into a natural brick called scoria. Over time the softer earth eroded, leaving the towering scoria behind.

Teddy loved this land and used to hunt in North Dakota. He said one reason he became president was to pass conservation laws to preserve these lands. T.R. created five national parks, fifty-one bird sanctuaries, four game preserves, and 150 national forests. Good on ya, Teddy!

Two days later, I arrived at my primary destination for the trip, Glacier National Park. Seventy-five million years ago the Earth's crust rose, forming the northwest mountains. Cold temperatures created snowfields and ice that over time became glaciers. Ice lying against the mountains would thaw and refreeze, plucking rocks from the mountains, causing it to erode to a toothlike horn. Deep glacier lakes fill the bottoms of the glacier valleys.

Glacier National Park, 2013

In 1806 the Blackfeet tribe sold Glacier property to the U.S. government for $1,500,000. The national park became established in 1910. It covers 1.4 million acres and still has glaciers, thirty-seven of them, to be exact; however, they are disappearing fast. Grinnell Glacier is one of the largest, estimated to be five hundred feet thick in places. In 1901 the glacier measured 525 acres but now is smaller than 300 acres.

Most scientists believe if weather patterns stay the same, the park's glaciers will be gone by 2030.

Glacier National Park is huge, and to see it you must hike on foot. There is but one road, called "Going to the Sun Road," and it climbs to 6,646 feet above sea level. It's fifty-two miles long and was completed in 1932. Construction of the road was an incredible undertaking. Workers had to be lowered on ropes in many areas to set dynamite charges in the rock face. There are two tunnels. It took ten years to complete the construction, because blizzards and snow permitted only four workable months each year. The snow closes the road at the end of October each year, with its earliest opening date recorded as May 16, 1987.

Traffic moves slowly and there would be few opportunities to pass other vehicles. It would be a full day's ride to go from one end to the other and return, especially if you stopped at the various parking areas to take pictures, and who could resist that? Numerous hiking trails exist at various points along the road, all well marked. My plan was to ride the entire route the next day, with occasional stops, of course, and return the following day to take a hike.

That night in my tent I read about the building of the road and eagerly looked forward to seeing the steep incline they described at Logan Pass, the highest elevation. The next day, after riding just a few miles and seeing the mountain ahead, I thought, Surely I won't be climbing that one. And then I saw the glint of windshields crawling across the mountain's face thousands of feet above me. *Oh boy, this will be fun!*

In a flash of brilliance, I remembered a rig I came up with back home for such an opportunity. I had purchased a small video camera on a previous trip but found it awkward to hold the camera and control the motorcycle at the same time. With that in mind, I designed a simple mount at the top of my windshield to hold the camera. I could easily start and stop the filming and after trying it out and seeing the images, knew it would be a winner.

My video of Going to the Sun Road the next day shows the hairpin turns, the 1,000-foot drop-offs, and the narrow roadway. Pickup trucks with large sideview mirrors often had to wait for another large truck to pass to clear the rock walls. My camera attracted a lot of attention. Approaching motorcycles could see what I was doing and waved enthusiastically. At a rest stop an old timer curious about the camera approached me. "Step in front of the bike," I said, and turned on the camera and interviewed him. He loved it.

I followed one of the red tour buses along the route. The red buses are famous at Glacier. More limousine than bus, each vehicle seats sixteen passengers and has an open roof. Whenever a red bus passed an attractive vista, the passengers would stand up, taking pictures through the open top. In 1936 the park bought 702 of the vehicles from the White Motor Company of Cleveland, Ohio. Those 1936 models are the same buses used today, with some modifications. In 2002 Ford Motor Company rebuilt all the old White vehicles, keeping the bodies the same but giving them a stronger, more lightweight chassis and new interiors, modern instrumentation, and engines that run on propane fuel.

Tour bus, Glacier National Park, 2013

I returned to Glacier Village later that afternoon and went to the store (also a restaurant) to buy ice. Well, why not eat something and not bother with dinner later? While sitting in the restaurant booth, I became aware of how grungy I appeared. My clothes were filthy, my hair was matted, my fingernails were all broken, and since the campground had no showers, good heavens, what must I smell like?

Once back at my campsite, just a few yards from Lake McDonald, I could see kayakers and water-boarders enjoying the waterfront. Since it was an unusually warm day, I decided to go swimming and wash off my grunge. Have you ever gone swimming in a glacier-fed lake? Yowie! It sure doesn't take long to get wet, I'll tell you.

My campsite neighbors, PZ, his wife Vicki, and son Dakota, invited me over for a bratwurst. They are from Spokane, and this was their annual Glacier campout. Every year they hike different trails. This day they hiked thirteen miles, and that is at six thousand feet, mind you. Dakota is a high school golfer (2 handicap) and a runner, so understandably, we talked golf and my son Brian. What a nice family ... and it was a great brat!

On my third day in Glacier, I did some hiking. To see 90 percent of this National Park, that's the only way to do it. Unlike Yellowstone, with hundreds of miles of roads, Glacier you do on foot. When the park first opened in 1913, the Great Northern Railroad put in a rail line carrying tourists to the park. It also built lodges for accommodations. When the visitors got off the train, it took a full day's horseback ride to reach the nearest lodge. I'm impressed when I realize that thousands of visitors signed up for that experience back in 1913.

I rode to Logan's Pass, where I found the trail to Hidden Glacier Lake. A park guidebook described the trail as a "moderate" mile-and-a-half-long walk with a gentle seven-hundred-foot climb. It couldn't be too arduous, I thought, since many of the other hikers were seniors, young children, and parents pushing strollers.

Whoa, I must be getting old. After fifteen minutes, I panted like a plow horse. All those other hikers passed me like they were going downhill. Must be the elevation, I thought. After all, we're at seven thousand feet. That must be it. They're acclimated, and I'm not. At least that was my excuse.

I paced myself and rested frequently while I enjoyed the scenery and watched numerous mountain goats roaming the snow fields. Back in 1920 this side of the mountain was under a hundred-foot-thick glacier. Now that glacier is all but gone, leaving behind only rocks and gravel that had been trapped in the ice for hundreds of years. After I reached the top, I looked down on Hidden Glacier Lake, described as being over a thousand feet deep. The hike back to the road was easier.

The campground at Glacier is primitive: no showers, no water at the sites, and no electricity. In order to find wi-fi and electricity for charging laptop, phone, and camera, I needed to ride into Glacier Village, where a photo studio allowed the use of its internet service but charged five dollars an hour to plug into electrical outlets. The studio was one of those shops where tourists dress up in cowboy outfits and have their picture taken standing by a split-rail fence.

It reminded me of the time my three sons — all grade-schoolers — and I camped in the Wisconsin Dells. The town had a photo studio, and we decided to have a picture taken. The young woman taking the order and helping with costume selection happened to be deaf but read lips and spoke reasonably well. "Do you want to be cavalry soldiers or cowboys?" she asked.

"No," I replied. "We want to be dance hall girls." That challenged her lip-reading skills, and had me repeat my statement several times. Finally, she understood and got in on the fun, bringing out hoop skirts, feather boas, and parasols. We attracted attention. Soon the photo shop windows filled with the faces of dozens of curious lookers. For the rest of the day, as my sons and I walked the streets of the Dells, people would point and say, "There's that father and his dance hall sons."

I would do another hike on my last day in Glacier, this one to Avalanche Lake. Although only a five-hundred-foot climb, it would be a five-mile round-trip, more than I'm used to walking for exercise. The elevated climb was gentle, and feeling proud of myself for my brisk pace, I could hear footsteps rapidly approaching from behind. A young couple appeared. A woman had a baby in her backpack, and a man carried a fifty-pound preschooler on his shoulders. They caught up to me like I had been standing still. Julia and Jonathan lived in Colorado, so this hike was most likely an everyday walk in the park for them. Tanned and trim, both mom and dad looked like they could have posed for the cover of *Outdoor Life Magazine*. I took their picture for my collection and had a pleasant visit. However, not wanting to interrupt their hike further, I said good-bye, and within minutes they were out of sight.

Keep it up, Julia and Jonathan. Enjoy these fleeting years. In a blink of an eye, you too will be in your mid-seventies.

What a beautiful walk. In the lower areas, the path followed a raging stream of crashing torrents of water. The roar of the water was deafening as it dashed against the boulders and threw spray on the stream banks. Farther up the trail, I entered a pine forest, weaving among trees that reached up two hundred feet and created a sunless tunnel. In another area, moss covered the entire hillside, giving it a soft, velvety appearance.

After a high point the trail, it then descended to Avalanche Lake, a glacier lake fed by two waterfalls. Some children splashed and swam in the ice-cold waters. I found a log and rested my legs, preparing myself for the return walk.

My time in Glacier had been delightful. I tried to compare it to Yellowstone and Grand Teton, but I find each different in wondrous ways. Yellowstone is huge and has a great variety of topography, but crowds of vacationers are everywhere. Grand Teton has those breathtaking mountains and lakes and is not so busy. Glacier National Park, on the other hand, hides 90 percent of its scenic wonders behind its mountains and forests. It's an explorer's wonderland but requires a hiker's determination to unlock its secret beauty. It had been an unforgettable week.

#

Chapter Thirty-Three

Travel Highlights

Besides the attractions already discussed in this memoir, a number of other travel highlights I had over the years while motorcycle/camping are worthy of a mention.

Devil's Tower, Wyoming

Devil's Tower, its rock columns extending 867 feet in the sky surrounded by flat prairie, is an improbable sight when initially coming into view. It was first formed underground fifty million years ago. The columns seen today were molded from molten magma. They were pushed to the surface along with the western mountains. Erosion of the surrounding soft materials left the arresting sight, marveled at by millions of tourists.

I visited Devil's Tower during a western motorcycle trip in 2013. When arriving on the site, I noticed the curious appearance of ants crawling up the side of the tower. Then I realized those weren't ants, but climbers who think it is great fun to wedge their hands and feet into the cracks between the columns and ascend into the lofty reaches.

You must register with the rangers before you climb, and because sudden storms are common, climbers must carry rain gear and a flashlight. Rock falls are also common, and helmets are advised. I was content to admire the wondrous sight from ground level.

Auburn, Cord Museum, Auburn, Indiana

In 2014, when heading north, eventually riding into Canada, I passed through Fairmount, Indiana, birthplace of the movie idol, James Dean. Back in my working days, whenever our company held a customer meeting, I would warm up the crowd by asking a trivia question: "In the popular movie *Rebel Without a Cause*, starring James Dean and Natalie Wood, what was Dean's character's name?" The answer is "Jim Stark." I realized I was getting old when I posed that question and people asked, "Who's James Dean?"

Just up the road, in Auburn, Indiana, is the Auburn Cord Museum. It is found in the original Auburn Factory showroom with its big showroom windows, chandeliers, and terrazzo floor. The cars are stunningly beautiful. Their huge chrome headlights, elegantly long hoods, and plush interiors made it easy to envision 1930s movie stars and gangsters wheeling down the boulevards in regal splendor.

The Auburn factory was originally a buggy manufacturer. However, with the growing popularity of the horseless carriage, the owners, the Eckhart family, switched from buggies to automobiles in 1907. The Auburn car had some success, but after the company hired a general manager by the name of Mr. E. L. Cord, the business really took off. Cord designed a car bearing his name that became a sensational hit. The company later purchased the Duesenberg Company of Indianapolis and sales skyrocketed even more.

The first floor of the museum is only Auburn, Cord, and Duesenberg models, but on the upper floor you can see cars made by Pierce Arrow, Rolls Royce, and Stutz, along with many other classic automobiles. What a great visit.

Mount Rushmore, South Dakota

Who are the four presidents carved into Mount Rushmore? Answer: Washington, Jefferson, Lincoln, and Theodore Roosevelt. Why those four? Answer: Because they represent, respectively, the birth of a nation, expansion (Louisiana Purchase), preservation (saving the nation), and development (Panama Canal and National Parks). Who is the sculptor? Answer: Gutzon Borglum. How long did it take him? Answer: From 1927 until 1941. What primary carving tool did he use? Answer: Dynamite. How did Mt. Rushmore get its name? Answer: Rushmore was an explorer on an expedition in the area in 1885 and named the mountain after himself.

The nighttime lighting ceremony is a spectacular show. The moving presentation focuses on the four presidents, their patriotism, and the nation's history. It's a stirring production.

Montreal, Canada

First explored and controlled by the French in 1534, this part of Canada was first known as Mount Royal. It remained under French rule until the Treaty of Paris in 1763, when all of Canada came under British rule. Surprisingly, the Brits were benevolent monarchs and accepted the French culture, guaranteed the use of the French language, and allowed the Roman Catholic religion. Nevertheless, two parallel communities existed in Montreal. Each group knew it was a vulnerable minority — the French in Canada, and the English in Montréal. Squabbling persisted.

While on my 2014 Canadian motorcycle trip, I camped in a lovely KOA campground outside Montreal and then took a Gray Line bus tour of the city. My first impression reminded me of the traffic in Paris. Thank God, I hadn't tried to ride my motorcycle through those chaotic streets.

Montreal is an island surrounded by the St. Lawrence River. It is like Manhattan Island in New York in that regard but is nine times larger. In addition to the traffic reminding me of Paris, high-fashion ladies are seen on the streets shopping the many French designers' shops. Also as in Paris, young people may have rings and things protruding from various places on their bodies. Beggars also populate the streets, cups in hand and children in tow. Just like Paris.

Another similarity is the high price of everything. Five Canadian dollars, nearly equal to American dollars, seems to be the starting price for any admission or trinket, no matter how small or trivial. The two-dollar Canadian coin (the toonie) is almost a joke, regarding what it will buy.

First stop on the Gray Line tour was the Basilica of Notre Dame, the largest Catholic church in North America. There are three tiers of balconies above the sanctuary. Celine Dion was married there. Anyone else wishing to have their nuptials performed in that church must put in their reservations two years in advance. Young, unspoken-for ladies are reported to make advance reservations, just to be ready.

The Pointe-à-Calliere, is the location of the Montréal Museum of Archeology and History. The museum was built on top of the remains of Ville-Marie, the earliest European colony. In the lower level of the museum, remains of the old city, its cemetery, and early sewer system are still visible.

Tribes of the Iroquois and Huron Indians first lived on the island from 1000 to 1600. Jacques Cartier, the French explorer, discovered the island in 1534. He was looking for a shortcut to the Orient at the time. Cartier named it Mount Royal because of its prominent elevation. Samuel de Champlain arrived in 1611 and established a fur-trading post. That didn't sit well with the Native Iroquois tribes, and fortifications needed to be built around the post for protection from the Indians. The walls now define an area known as Old Montréal.

The British in the 1700s sought a foothold in North America, and that led to fighting between the French and English that did not subside until the English victory and the Treaty of Paris in 1763. There remains a cool relationship between English and French to this day.

The rest of the tour involved visiting the Latin Quarter, seeing the Museum of Fine Arts, and strolling about the lofty heights of Mount-Royal Park overlooking the city.

Quebec, Canada

A day after leaving Montreal, I arrived in Quebec, ready to learn its history. I found it quite similar to Montreal's in many respects. Champlain established a fur-trading post here, just three years after the one he started in Montreal. Quebec was situated on cliffs even higher than the elevation at Montreal. The cliffs protected Quebec from attacks by the British in the 1700s. However, in 1759 the British finally succeeded in summiting the cliffs, getting behind the fortified city and capturing it.

Quebec, again like Montreal, is completely surrounded by a wall. After the establishment of peace in 1763, they considered tearing down the wall but then realized it was a major tourist attraction, and the wall remains standing.

Other "must see" attractions in Quebec include Old Quebec, the Royal Palace, and the Citadel. The streets, as in Montreal and Paris, are a tangle of motor cars, multiple lanes of honking, speeding vehicles, distracted jay-walking pedestrians, and undecipherable signage. How fortunate I was to again have taken a shuttle into the city and not try to ride my motorcycle.

Leaving Quebec the following day, and looking for the bridge over the St. Lawrence that I was told I couldn't miss, I made a wrong turn and ended up deep in the heart of Old Quebec. Oh, no. My worst nightmare! Not trusting myself to avoid one-way streets or making wrong turns, I tucked in behind a taxi and followed it blindly, regardless of where it was headed. Fortunately, it led me out of downtown. I asked a street worker for directions. Even though he did not speak English, he pulled out a smart phone and showed me a map of the route leading to the U.S.A.

Crazy Horse Monument, South Dakota

When visiting the Dakotas in 2013, I wasn't sure I wanted to bother stopping to see the Crazy Horse Monument. After all, only the head of the monument had been completed before the artist's death in 1982, and even though work continued, it would be another sixty to sixty-five years before it was finished.

However, it wasn't far away, and at the most, I thought I would do a thirty-minute drive-by and then be on my way. I ended up spending an entire day at this fascinating landmark.

Mount Rushmore, only a few miles away, was finished in 1941 and received much praise for its promotion of patriotism, freedom, and human rights. Chief Henry Standing Bear, of South Dakota, contacted an artist just becoming known at the time, Korczak Ziolkowski, saying, "My fellow chiefs would like the White Man to know the Red Man has great heroes, also," referring to Mount Rushmore. No money was offered, but Ziolkowski took on the project with his own financing and what money he could raise. The federal government offered several times to help with funding, but Ziolkowski and the Indians refused all offers. The Native Americans' distrust of government, after so many broken promises and violated treaties, led them to decide the project would be completed without government help.

Crazy Horse is best known for his defeat of Custer at Little Big Horn. Although he was a fearless warrior, most of his raids were against hostile Indian tribes. Once the white man started moving into Indian Territory, his focus changed. Almost all of Crazy Horse's attacks on settlers were in retaliation for Indian women being kidnapped or braves being killed. Little Big Horn was fought on land seized by General Custer that had been given to the Lakota tribe by treaty in 1868. Crazy Horse was shot and killed at Fort Robinson in 1877 while negotiating with cavalry troops under a flag of truce.

The sculpture of Crazy Horse is huge. All four presidents at Mount Rushmore could fit inside the head of Crazy Horse. The park is awesome. Not only do they have viewing verandas, but there are movie theaters, conference centers, gift shops, museums, sculpture workshops, and restaurants. I left impressed by the Native Americans' culture and shamed by our treatment of them.

Bath Shipyard, Bath, Maine

Bath Shipyard has been building ships for 125 years. In 1893 three companies were building wooden sailing ships in Bath. When industry changed to steel, ship builders found their employees in the engine and boiled trades. The most recent notable ship built in Bath was the *Zumwalt* in 2014, a guided missile destroyer built of exotic materials undetectable by conventional radar.

Ships are no longer built on slideways that slide the ships into the water when construction is completed. Today's ships are built in sections; bow, middle, and stern, on level ground, then joined and transported to the water in moveable dry docks.

The museum at the shipyard has artifacts of sailing days such as seaman's bags, masters trunks, and sail-mending tools, as well as paintings and models of exciting seafaring history. My two-hour allotment for the museum visit was not enough.

Badlands National Park, South Dakota

I can't find the words to describe the Badlands. Frank Lloyd Wright in 1953 struggled as well. He said, "I was totally unprepared for that revelation called the Dakota Badlands. What I saw was … an endless supernatural world more spiritual than earth but created out of it."

Even though I was eager to get to my campground that day, I couldn't keep from stopping to admire and take pictures of the incredible sights. The peaks and valleys were made up of a blend of delicately banded colors. It was dramatically barren, eroded earth, with high buttes and deep gulches, and nary a blade of grass to be seen. It was like another world.

Eastman House, Rochester, New York

George Eastman, founder of Kodak Camera, was a wealthy and generous benefactor to numerous worldwide causes. Born in 1854 to an ordinary family, his father died young, and his mother, a widow at age forty-one, took in boarders. George quit school at age thirteen and earned money as a newsboy to contribute to the family welfare.

Photography in the late 1800s was available only through professional photographers. Using a plate system, the photographer put himself under a drape, snapping portrait pictures with a camera mounted on a tripod. A friend suggested George learn the photography business. George was an intelligent individual and contemplated converting the plate system to a film, and putting it into a portable device that the general public might use. Would there be a market for that? he wondered.

Eastman called his company Kodak for no other reason than that he liked the sound of "K." The cameras were expensive by 1890 standards, costing twenty-five dollars. The camera came in a wooden box with film enough to take one hundred photographs. Once the pictures were taken, the camera was put back into the box and shipped to the Eastman Dry Plate & Film Company, along with ten-dollars for developing. The business boomed.

The next innovation was removable film, reducing shipping costs. Then came the "Brownie" camera, so simple a child could use it. Brownie cameras sold for one-dollar. Sixty thousand employees later, Eastman Kodak figured it had a successful, permanently established business. Kodak had patents for digital technology but sold them, thinking film would be around forever. It resulted in the eventual bankruptcy of Eastman Kodak.

Back in Kodak's heyday of 1905, George built a house. The home had nine fireplaces, fifteen bathrooms, eleven bathtubs, and a turnstile in the garage, so you wouldn't have to back the car out. The eight and half acres around the house had chickens, cows, horses, and a vast vegetable garden. Inside his conservancy was the largest pipe organ known to exist in a private residence, with over a thousand pipes. At one point Eastman's house was cut in half and dragged nine feet apart to allow further expansion. The house had an elevator, because George's mother lived with him and was doing poorly. She used a wheelchair and had no teeth. Because of his mother's dental problems, Eastman started five dental institutes around the world.

Eastman was a rather vigorous individual who enjoyed hunting and fishing in the wilds of America. Unfortunately, by middle age, he suffered from diabetes and debilitating arthritis. At age seventy-eight he rewrote his will, detailing how his wealth would be divided, and scribbled the following note: "To my friends, my work is done. Why wait. Signed GE." He shot himself in the heart with a revolver. What a sad ending for such a generous and interesting man.

I mentioned earlier that a friend suggested I name my motorcycle Rocinante after Don Quixote's horse. It was a great name.

A second choice might have been Serendipity, the aptitude for making desirable discoveries by accident. That's how I think of these unstructured travels of mine. Incredible discoveries that I find almost by accident as I travel across the country, marveling at the sights and the history I learn. How lucky I have been to have had these opportunities.

###

Chapter Thirty-Four

Sailing the Greek Isles

Whenever experienced sailors talk about their favorite sailing playgrounds, the Greek isles are always mentioned. They rank right up there with Tahiti, Bali, and the Caribbean islands. Greece is estimated to have six thousand islands and islets. Beaches stretching over many kilometers, sheltered bays and coves, coastal caves, and blue-green crystal-clear waters distinguish this cruising paradise. To sail the Aegean and Ionian Seas is an extraordinary experience.

I fantasized about such a destination and late in 1999 contacted *Cruising World* magazine, which I knew sponsored adventure flotillas to exotic parts of the world. Sure enough, the magazine had scheduled a trip to the Aegean islands for April 2000, during the Greek Easter celebrations. Orthodox Greeks celebrate Easter with gala tradition and festivity. Their Easter observance is bigger than Christmas.

We asked our Bloomington friends Dave and Terry Baer to consider joining us for the trip. Dave retired as president of one of the Bloomington banks a year ago, and he and Terry had been traveling frequently since then. The Baers are sailors, having owned a sailboat on Lake Monroe, and also have taken sailing charters.

Dave is also a former Navy pilot, which gives the two of us a lot of the exercise talking with our hands. "There I was upside down at 10,000 feet ..."

The flotilla would include twenty-six other paying tourists sailing four yachts, fifty to sixty feet in length.

Travelers would be expected to crew the boats, but each would have a Greek captain aboard. The managing editor of *Cruising World*, Lydia Childress, would also be one of our sailing companions on the trip. Lydia, an attractive young lady in her thirties, had sailed in the Aegean many times. She claimed to spend at least one month there every year. We learned later that Lydia is engaged to one of the Greek captains, which explained her attraction to this part of the world.

Cruising World used to have a full-time person on staff to orchestrate their adventure tours. It no longer had such an employee and instead contracted with King Yacht Charters to handle the arrangements. King engaged GPSC (Greek Private Sailing Charters) to provide services for the cruise. GPSC was one of the largest charter companies in the Aegean. It operated out of the Athens harbor marina along with Moorings and other charter companies.

Unlike dealing with one organization, making arrangements through three companies created confusion and conflicting information. Those who had sailed with *Cruising World* before said planning this trip was not as flawless as past adventures. Make the best of it, we told ourselves, and don't let it spoil the experience.

From Indianapolis we flew through eight time zones getting to Greece. We left Indy at 10:40 a.m. Wednesday, and with one flight change in New York, we arrived in Athens at 9:00 a.m. Thursday. Our twenty-six sailing companions were also on the New York flight, although we didn't know who they were. We would meet them for the first time at a welcoming cocktail reception at Athens' Royal Olympic Hotel, where we would spend our first night.

We met at 7:00 p.m. to socialize and learn our boat assignments. Aboard the lead boat, *Alisson*, a sixty-five-foot sloop, would be Vince and Pat from Charleston, and Ted and Nancy from San Diego. The two couples are friends from the Midwest, now living in different parts of the U.S. The couples met as undergrads at the University of Michigan and have remained close for the past forty years.

Margaret, a widow from Providence, Rhode Island, would also be aboard their boat. She was about sixty-five and had a genial look about her that invited conversation.

"Margaret, we're Jim and Michele. Have you been on cruises like this before?"

"Oh, yes. Many times. In fact I returned two weeks ago from sailing with the Moorings in Tonga."

"Really! How did that go?"

"Delightfully. Although the boat was skippered by a captain, I own it and lease it back to the Moorings Charter Company. After my husband died, I sold our fifty-foot Beneteau in the Chesapeake and sneak out to the South Seas at least twice a year. Tonga is known as the 'friendly Islands' for their cordial reception of Captain Cook back in 1773. The natives there are wonderful."

"Margaret, we'll have to spend an afternoon visiting. We want to hear more about that."-

The Starks and the Baers had been assigned to the *Frantzeska*, a sixty-one-foot Venus 18. In addition to ourselves, there would be Wayne and Nan from Tucson, and Steve and Marian from Seattle. Both Wayne and Nan are retired college professors, and Steve is a software developer.

Wayne and Nan were a sweet couple but quiet and shy. Their answers to our questions about their sailing experiences and other personal inquiries were especially brief.

They taught at various colleges over the years. His field, biology. Yes, they sail. Yes, they charter. They once lived in Indiana.

"Oh, where?"

"Terre Haute."

"Did you like it?"

"No."

Quiet conversations on the deck with Wayne and Nan over the next two weeks would be … well, quiet.

Steve and Marian were interesting. Married? Don't know. She was close to six feet tall, and he was five-four, at best. Most people at the cocktail party had slicked up to some degree, given the casual nature of packing for two weeks on a sailboat. Steve had on rumpled khakis, a matching long-sleeved work shirt, with one sleeve rolled to the elbow, the other hanging open-cuffed below his fingertips. He wore flip-flops on his bare feet. Steve didn't mingle, just hovered over the hors d'oeuvres table like it was his last meal.

Marian's voice carried above all other sounds in the room. It was a screech, like opening a large barn door. "Steve! Steve! Save some for the others!"

Charless (yes, two sses) and his wife, Marjorie, according to their nametags, were from Dallas. His belt buckle should have given me a clue.

"What do you do in Dallas, Charless?" I asked.

"I sell geoexchange energy systems."

"What the heck is that?"

The minute I asked, I sensed my mistake. Charless was off and running.

"It uses the earth as a heat source or a heat sink. Depending on latitude, the temperature beneath the upper six meters of Earth's surface maintains a nearly constant temperature between ten and sixteen degrees Celsius. Ground-source heat pumps extract heat in the winter ..."

"Oh, excuse me for interrupting, Charless. This is my friend Dave. Dave, Charles was just telling me about his interesting business. Oh, I see Michele is giving me that 'come hither' signal. Better go see what she wants. Excuse me."

Dave still reminds me of that dirty trick.

Adonis, the tour guide for the entire group, would also be on our boat. Adonis welcomed the group.

"Friends, we are so pleased to have you with us for the next ten days. Tomorrow, we will leave the hotel at nine sharp. Please be ready to go in the lobby. The bus will take us to the harbor, where we will board a hydrofoil for a two-hour, high-speed transit to the island of Hydra. There you will go aboard your assigned boats. If you miss the bus tomorrow morning, it's a long swim to Hydra.

I have prepared handouts showing the islands we will visit over the next ten days. There is also a list of the names of all guests, plus the crew.

You're on your own for dinner tonight, but I recommend any of the fine restaurants in the Plaka, just a block from the hotel. Welcome to Greece, everyone."

Hydrofoil out of Athens Harbor, Greece, 1999

The hydrofoil harbor the next morning reminded me of the Greyhound bus station. It's a major departure point, and patrons swarm in bunches to get aboard one of the departing boats. The hydrofoils themselves look like something out of a Jules Verne novel. The torpedo-like shapes reminded me more of a submarine than vessels that skim a few feet above the water. They might look like submarines, but riding a hydrofoil is more like being in an airplane. Passengers sit in airliner-type seats. Almost 90 percent of the vessel is enclosed. There's an opening amidships and one in the stern, but it is small. Because of the noise of the powerful engines, only those dying for a cigarette or already half deaf can stand it. Since I fit into the latter category, I was able to spend the last half of the trip in the open air. The ride itself was smooth. The hydrofoil rises above the waves on ski-like appendages, with only its props extending into the water. I estimated the hydrofoil to be traveling at thirty knots.

The island of Hydra must be the place photographers use for all those calendar shots of the Greek isles. Turquoise waters are encircled by hillsides dotted with clinging white villas. The harbor itself is a cornucopia of waterfront shops, open-air restaurants, village churches, cobblestone streets, donkeys, pushcarts, street vendors, sun umbrellas, and boats, boats, boats.

There are fishing boats, sailboats, work boats, passenger ships, ferries, and luxury yachts. Periodically a church bell would sound a single mournful Good Friday peal.

Our arrival in Hydra was chaotic. "Chaos" is a fitting

word, but only in the sense of energy and excitement. Anxiety was not one of our emotions, since at that point we still had confidence in GPSC. We had but to await their instructions and move off in the instructed direction.

Harbor master, Hydra Island Harbor, Greece

The Hydra harbor master was the personification of a fireplug. His less-than-five-foot frame was supported on treelike legs connected to a barrel-shaped torso. From this barrel sprouted two powerful, hair-covered arms. His full beard almost hid his massive chest. He arrived at the quay-side in a puckishly colored red and blue dinghy to take our bags on a shortcut from one side of the harbor to our waiting boats on the other. We walked around.

After dumping our bags in our stateroom — a grand term for a five-by-eight-foot space with adjoining head — we walked back into the town's center for provisions. The boats have no supplies other than bedding, two towels, cooking utensils, glasses, cups, and silverware. Luxury items such as napkins, paper towels, toilet paper, and matches to light the gas stove would be up to us to provide. We returned to the boat a short time later with wine, coffee, snacks, and some of the abovementioned luxury items.

Adonis, our genial guide, who had been with us since Athens, gathered us together at 5:30 for a tour of the town. "Hydra," meaning "water," actually refers to the fact Hydra has *no* water. Freshwater is brought onto the island each day by ship. The island was originally settled by inhabitants from the northern regions of Greece. Not understanding the needs of the area, they built their homes with peaked roofs, just like those they had left in northern Greece. Flat roofs to catch rainwater would have been so much more utilitarian.

We visited a small monastery in downtown Hydra and were amazed at the ornate nature of its churches. This was our first look inside a Greek church, and we would discover over the next two weeks that this is the nature of all churches in this country. Orthodox traditions go back many centuries. Therefore, when one of the priests walked out of the church holding a cell phone next to his ear, we found it somewhat incongruous.

Walking up the surrounding hillside, we were given a spectacular view of the harbor down below. The sight of heavily loaded donkeys, simple pristine-white villas, old windmills, and colorful gardens transported you to another time and place. But then the discovery of Disco Heaven amid this antiquity brought the twenty-first century roaring back.

Good Friday is a solemn day in Greek Easter observances. All throughout the day, that single sad clang of the church bell would sound, noting the crucifixion. We gathered at 8:30 p.m. to walk to the nearby village of Kamini. There we would watch the Epitaphios parade.

Epitaphios, meaning "epitaph," is the symbolic funeral of Christ. After dark, a candle-lit parade, led by clergy bearing flower-draped biers, snakes through the various Greek villages and ends at the sea. On Hydra the Epitaphios ends at Kamini, where the bier is finally walked into the water.

A table for dinner had been reserved for the twenty-six of us on the waterfront. Nevertheless, with thousands assembled to see this spectacle, we had to stand on our chairs to catch a glimpse of the parade over their heads. I raised my camera at arm's length and blindly clicked off several shots.

Our dinner, lasting close to two hours, included traditional dishes of fried squid, tiny grilled fish, and lima beans in sauce. Really quite good. It was a late night.

On waking the next morning, I walked around the quay, found a table at one of the many sidewalk restaurants, and watched the town come to life as I drank my heart-starting Greek coffee. Easter Saturday is a day of fasting for most Greeks. That night, religious services begin, leading up to great celebrations at midnight. Fireworks and gala banquets announce the celebration of Christ's rising, and we were to participate. Our evening would not start until 11:00 p.m., so pacing oneself would be important that day.

The traditional celebratory meal that evening is lamb fixed on a spit. As I drank my coffee on the waterfront, I saw a number of burros delivering sheep carcasses to small boats. Each is wrapped in a black plastic garbage bag. It took some deciphering to figure out what was in the bag. The protruding skinned lamb's head with its surprisingly large eyeballs finally decoded the bundles.

What an experience it is to be in this setting, surrounded by sights and sounds so unique to normal day-to-day activities. I loved it. The local citizens almost all wear dark, heavy clothing. Although I was comfortable in my T-shirt, the donkey drivers all wore long-sleeved shirts under sweaters, and many had coats or jackets over them. Women were clad in long black dresses, with sweaters and heavy dark shoes. All seemed stoic and stern in appearance, but most responded willingly when requests for service were made.

Soon my GPSC boating companions started to appear, stumbling out of the grogginess of their late-night revelry. Most bought pastry from one of the nearby bakeries and then, like me, enjoyed the morning at one of the wharf-side tables. I bought a croissant for Michele to headed back to the boat.

At mid-morning Michele, David, Terry, and I left to explore the island. Big passenger boats and ferries were arriving from Athens at that hour. There were thousands of visitors swarming over the quaint village. Some boats had identified their groups by giving them different-colored hats. Small pockets of multinational tourists were bobbing about in red, yellow, and green beanies.

One end of the quay led up to a fortlike structure, complete with cannon and brilliantly hued flags. All tourists were taking photos of companions standing on, hanging from, or leaning it against various battlements. Those still using film cameras made Kodak and Fuji exceedingly happy.

As we worked our way up the surrounding hillside, we broke away from the crowd and found ourselves alone, wandering among the lovely white hillside villas. Dave suggested he could probably spend several weeks in one of these elevated retreats with only a stack of books to keep him entertained. I agreed, but I would want a sailboat tied in the harbor for occasional escapes.

The gardens, walkways, walls, and villas were covered with bougainvillea, roses, poppies, and blossoms of unknown names. With no motor vehicles on the island, we wondered how villa owners lugged their goods and supplies up to these lofty habitations. Then the size of the droppings on the walkways revealed the utility of the donkeys seen along the waterfront. Every new level unveiled another photograph begging to be taken. I went through several rolls of film.

On return to sea level, the four of us had a leisurely lunch, then shopped for additional provisions before returning to the boat.

Back at the *Frantzeska* we learned a problem had been discovered. Our boat was not able to heat water. Living in close quarters for ten days would be one thing, but doing it without a hot shower would be a whole other matter. All four captains were wrestling with the problem, testing various systems, trying to identify the cause of the malfunction. The twin diesel engines were running during this trouble shooting, making my plans for an afternoon siesta impossible.

I retreated to one of the neighboring boats to read. Dave ignored the noise, diesel smoke, and Greek discussions and curled up in his cabin. The ladies went shopping.

The quay to which we were tied was five boats deep. Our four boats were clustered together, two against the sea wall and two tied to their bows. Three other boats extended out into the harbor beyond our boats. It is an acceptable practice when entering a crowded harbor to tie up to boats already there and walk from one to the other until reaching shore. All boats had dropped an anchor before backing into their berths in typical Mediterranean mooring style. The speculation about tangled anchor chains made departure tomorrow intriguing to imagine.

At 6:00 p.m. all the GPSC sailors began to return from their day's adventures and gather about the boats. Wine was uncorked and conversation generally focused on the *Frantzeska*'s hot-water problem. The captains explained they had exhausted all known quick fixes and had even called GPSC about a replacement boat. None was available, so they ordered a part that would arrive the next day, and, they assured us, that would surely correct the problem. For tonight, however, it would be a chilly shower for those so inclined.

Lynda Childress, our *Cruising World* hostess (did I mention she also happened to be young and attractive?) came over to our boat to apologize for the problem and offered the use of the shower on her boat. Michele claims my immediate and enthusiastic acceptance of the invitation (did I mention Lynda is very attractive?) was not entirely motivated by my interest in hygiene. Hey, not everyone can claim they shared a shower with the managing editor of *Cruising World* magazine (did I mention ... oh, yeah, I did).

As we congregated on the quay just prior to the 11:00 p.m. walk over the hill to Kamini, our captains started throwing firecrackers in advance of the midnight celebrations. The explosions were "cherry bomb" size and caused us all to be alert for hissing fuses. The captains, all young and loosened up by their ouzo, were not winning any points for professional decorum that night.

At 11:00 we left the rowdies on the quay and followed the path taken the previous night over the hill to Kamini. This time, however, we turned off when reaching the village and followed the throng to a small and very crowded village church. We squeezed inside and watched and listened to a rotund priest sing a mournful Easter Mass. At 11:50 the electric lights of the church and surrounding buildings went out, and candles were lighted.

Clamorous church bells announced the stroke of midnight, although barely audible over the exploding fireworks. In the brilliant orange of a thousand candles, we watched villagers kissing one another on both cheeks, both men and women and men and men. New Year's Eve could not have been more festive at that moment. It was Christmas and Fourth of July rolled into one.

Carried along by the joyful crowd, we worked our way to a village restaurant where a table had been set for the twenty-six of us for dinner. I found out later one of our members had stumbled exiting the church and suffered a sprained ankle.

The dinner was a traditional Easter meal starting with a soup made from sheep innards. As disgusting as that might sound, it was actually good. Roasted lamb and other more recognizable dishes followed. All were delicious.

It was 2:30 a.m. by the time we returned to our boats. It had been a long but fun day.

We were leaving the island of Hydra the next morning and sailing to the uninhabited island of Dorkas. The GPSC crew would be hosting a beach barbeque, preparing lamb on a spit along with all the traditional Easter foods.

I was looking forward to the departure, if for no other reason than to watch the before-mentioned untangling of the anchor chains. As ours would be one of the last boats to leave, I found myself a good seat on the bow to witness the show. When entering a harbor, a boat motors away from the quay seven or eight boat-lengths before dropping its anchor and backing up. In a small harbor that means the boat starts close to the opposite shore, laying its anchor chain forward of its bow as it backs up. In a crowded harbor, such as the one we were in, there must have been twenty to twenty-five anchor chains lying one on top of the other. If the boats were to leave in reverse order of the way they entered, last in–first out, it wouldn't be so bad. But that seldom happens.

The unscrambling technique, once you discover you have two or three chains hanging from your anchor's flukes, is to pass a line under the snagged chains, holding them high, then pulling your own anchor free of the snags and getting loose. It's rarely successful on the first try.

We watched amused as boat after boat unscrambled the spaghetti bowl. A power yacht in front of us snagged the chain of one of the GPSC boats and pulled it over its prop shaft. Although I assumed that was a disaster for sure, the inboard-outboard power boat was able to lift its prop, thereby reaching the fouled chain and getting it free. We could not tell if the screw had been damaged.

Although the winds were good, we motored the four hours to Dorkas because we had gotten such a late start in the day. I was disappointed to miss a potentially good day of sailing.

Turning into the inlet on Dorkas, we could see the lead boat, which had arrived hours earlier, had a shelter erected and a lamb turning on a spit.

The yacht *Frantzeska*, Greek Isles

The water was crystal clear in the inlet. Although its tropical appearance was reported to mask its cold temperature, it looked inviting. I knew a swim would be brief, but grabbing my swim mask, I dove off the bow and hurriedly swam for the shore. Not so bad. In fact, I swam out a few feet and made a couple of surface dives to check out the bottom. There are no coral or tropical fish in the Mediterranean; however, the sand and rocks were pretty, and my brief but invigorating plunge was enjoyable.

The meat was soon cut off the carcass, and each crew brought a platter back to be eaten in their boat's cockpit. Several side dishes were served along with the lamb. It was a delicious meal.

We were told we would be sailing out early the next day, but GPSC's credibility was starting to wane. At 9:30 a.m. we were still moored at Dorkas, and the irritations of the previous day were rekindling. The spark that set off the conflagration was Veronique's departure.

Veronique was the ranking GPSC representative. Her incredible lack of compassion for the problems of our boat had been remarkable. We kept expecting she would come over and express apologies about our difficulties. Therefore, when the water taxi pulled up that morning, and Veronique went zipping back to Athens, her "Have a nice day" was the last thing we wanted to hear.

Michele and Marian lunged at Adonis, the next GPSCer in charge. They unloaded on him with both barrels. It was more than just the water problem. The boat hadn't been cleaned very well, we should have expected at least some basic supplies on board — like toilet paper — and on top of everything else, GPSC seemed to consider this "their party at our expense." All weekend long, strangers kept popping up at our activities, invited by the GPSC folks. Adonis did seem concerned about the situation, and we also felt Lynda was embarrassed by our problems. We wanted to get beyond these dilemmas, since there were so many things we could be enjoying.

Crew of Frantzeska, L to R, Terry, Marian, Dave, Steve, Jim (helm), Michele, Nan, Wayne.

There was little wind when we left Dorkas, and we had no choice but to run under power as we headed toward our next port, Leonidio, on the Peloponnesian coast. After an hour a breeze could be felt, and we hoisted sail. The wind was light, between six and eleven knots, but we sailed for six hours, making between four and five knots. That's not an exhilarating speed, but I could feel my spirits lift just being under sail. We all enjoyed a pleasant lunch on deck.

A mile from our destination, George got a call on his cell phone that the other boats had already tied up, and they suggested that we start engines and join them. We were ready and motored in.

If anyone thinks the land of the ancient Greeks is not up to date with technology and wireless communication, they ought to see our Greek captains. Each carried a short-range walkie-talkie radio and had a cellular telephone. And they talked on them constantly. They used this communication legitimately to coordinate maneuvers when coming into port, but mostly their use was to taunt the other captains about sailing skills or to set up a close formation for the purpose of passing coffee or cigarettes ship to ship.

Lynda met us at the quay and announced she had made reservations for the crew of *Frantzeska* at a waterfront hotel. In addition, after we had a hot shower, she invited us for cocktails on the landing in front of our hotel. Excellent. If this had been the response when we first started having troubles, it would have been so much better.

The hot shower felt wonderful. After slipping into some fresh clothes, we met Lynda at a table at water's edge. The surrounding scenery was right out of a Greek travel brochure. The village consisted of a semicircle of shops, restaurants, and markets that wrapped themselves around the colorful blue harbor. Our four flotilla boats were tied Mediterranean style on one side of the harbor. On the other side were a variety of fishing boats, all decorated in their vivid red, green, and blue colors.

The small beach was littered with nets, buoys, and the typical flotsam and jetsam found in a village of this type. It might be trash anywhere else, but here it was acceptable ornamentation.

The *Frantzeska* crew had a pleasant visit with Lynda. We were pleased and appreciative of her efforts to soothe our ruffled feathers.

We ate our dinner in the same establishment in which we had our cocktails. Michele and Terry had met the lady who owned the restaurant, and they decided it would be a good choice. As we slid into our seats, Dave had been told we must go into the kitchen to select our food. They had no menus.

As I entered the kitchen, I swung the camera to my eye to capture the wall hangings of hundred-year-old sepia photographs, fake May Day flowers, and cut-out magazine advertisements. I then moved in for close-ups of the huge cast-iron stove and its bubbling, simmering, steaming pans of smelt, chicken, lamb, rice, meat balls, and shrimp. Michael, the waiter, stood by with a notepad, jotting down our selections. We all thoroughly enjoyed the choices we made.

Strolling along the crunchy pebble beach in the twilight after dinner, Terry searched the stones for interesting shapes. Some phallic discoveries led to speculation about the morals of some of our fellow passengers, and the conversation deteriorated from there.

It felt like we turned in early that night, but having had dinner at 9:30 p.m., it *was* approaching midnight. Besides, that night we had the luxury of a huge (standard double) hotel bed.

The next morning, we enjoyed one more hot shower before breakfast. The restaurant seemed deserted, but after wandering around I found two elderly folks in the kitchen. "Breakfast?" I asked. "Yes, please," came the reply. "How many?" he asked. "Four," I said. He then turned and busied himself with other things in the kitchen.

When I returned to the table, I told Terry and Michele that I didn't fully understand the conversation I just had, but I think I ordered breakfast. Sure enough, within minutes, an elderly gent returned bearing a tray of bacon, eggs, bread, jam, orange juice, and steaming coffee. Just what I would have ordered.

Adonis had arranged taxis for the entire GPSC crowd to visit a mountainside monastery. The taxis arrived promptly at 9:00. Because we would be visiting an active religious monastery, built some six hundred years ago, Adonis advised the men to wear long pants and the women to wear skirts.

We followed a beautiful winding passage through the mountains, mostly uninhabited. The view outside our Mercedes cab was of hairpin curves, lush green vegetation, and jagged rocky hillsides. After twenty-five minutes, the cab slowed and we were able to see high above us the angular dimensions of a manmade structure clinging to the side of the mountain. It appeared to extend out from the vertical mountain wall. How they were able to build such an edifice — particularly six hundred years ago — was everyone's first question.

Ten minutes later, we arrived at the cloister. We learned the convent was actually a nunnery. Ten nuns lived here to "do the Lord's work." Adonis explained that the popularity of the cloister as a tourist attraction provides more than an adequate income for its inhabitants. Indeed, the recently applied white wash, fresh carpentry, and generously decorated chapel suggested a very positive cash flow.

We stopped in the heart of Leonidio on the way back to the boat to do some provisioning. One intersection of this charming hamlet contained a food market, a produce stand, and a bakery—everything we needed. All but ice. The driver made several unproductive stops, until finally finding a vendor willing to give us the equivalent of an ice cube tray of ice cubes.

There are not good weather-forecasting services in the Aegean. Fortunately, our captain's father was a captain for the Moorings back in Athens. Because Moorings has its own forecasting services, George, our captain, called his father each night on the cell phone to find out what the weather looked like. We learned that a front would be moving into the area with high winds and high seas. It hadn't arrived yet, however, and we motored in calm winds toward the ancient city of Monemvassia, on the Peloponnesian coast.

Michele fixed lunch en route consisting of a Greek salad bathed in olive oil and a plate of bread, cheese, and bologna. All accompanied by a bottle of Retsina, a Greek wine that tastes like Chardonnay mixed with turpentine. How Greek can you get!

Monemvassia is called the Greek Gibraltar. First settled in 582 AD, the rocky peninsula broke away from the mainland in the fourth century AD during an earthquake. It does resemble that other famous "Rock." The rock became an important fortress because of its single opening from sea level up to its thousand-foot apex. The name Monemvassia means "single entrance" in Greek.

After mooring, Adonis announced that his tour of Old Towne would leave at 5:30. We saw the ruins on the hilltop high above us, but, not understanding the wonder of the village below the summit, I dressed for the walk in sunglasses, shorts, and T-shirt—assuming we would return to the boat before evening activities. The Olde Towne village was hidden from view from the quay. We walked fifteen minutes up the coastal road to the "single entrance." Passing through the entranceway, it reminded me of Mont St. Michele on the French coast. One look at Michele told me she was thinking the same thing.

The cobblestone passageways were narrow, so much so that in many places you touched buildings on both sides of the street by extending your arms. The homes, shops, and businesses in Monemvassia dated back to the fourteenth and fifteenth centuries.

The Venetians first established the fortress at the top of this rock in the sixth century AD. They built many of the structures that are still standing. Michele suggested we ought to take pictures of these 1,400-year-old walls to show to our builder's stone mason in Indiana. He has trouble keeping our condo's stone walls up beyond the ten-year homeowner warranty. We discovered Old Towne to be an enchanted village—fairy-tale-like—and we did not leave until later that night.

The plan the next day had been to assemble at 8:30 for a revisit to the top of Monemvassia to see the fortress. However, while we were still in bed at 6:00 a.m., movement of the boat indicated an increasing wind, announcing the arrival of the weather front we had talked about the day before.

The captains decided we should change our schedule and leave Monemvassia immediately. Our harbor was safe enough in case of a gale, but the open sea just outside the harbor might build ten-foot waves, trapping us inside for two or three days. If we escaped right away, we could head north into more sheltered waters, and even though the winds might be strong, the sea state would not be unnavigable. By 7:00 a.m. we were underway.

The wind blew twenty to twenty-five knots. We pulled out half the mainsail to stabilize the boat and then, later, more and more jib as conditions moderated somewhat. With lumpy seas and salt-spray in the air, most of our crew remained below deck. No one admitted to being ill, but several acknowledged moments of queasiness. At one point I went to our cabin to put on an extra sweater, and while moving about, opening cupboards and drawers, I experienced a touch of that same dizziness. Fresh air in the cockpit quickly revived me.

After several hours of hard beating into the wind, we shut off the engines and charged ahead under sail power alone.

The wind had dropped to fifteen knots, although the sea remained lumpy. The other boats were barely visible, doing their own thing on different headings and tacks.

Our destination was the island of Spetses. Approaching the island after several hours, the rain showers that we had skirted caught up with us. I was at the helm enjoying the macho exuberance of having the wind in my face, squinting through the driving rain, while steering the boat into six- foot swells. Although I had layered a sweatshirt over a sweater and had a jacket over that, I also wore shorts. That confounded those around me, who were bundled up from head to foot. I was actually comfortable. Another matter was the rolling, plunging, jerking motion of the boat. That was something I discovered I might not like for several days at a time.

With Spetses in sight, we dropped the sails and motored on in. The harbor had a number of boatyards. After tying up, Dave, Terry, Michele, and I toured the yards. Most boats under construction were small, twenty- to thirty-foot fishing boats. All had heavy keels, broad beams, and hand-shaped ribs. The nearby wood shops were primitive in their equipment and design but obviously got the job done the old-fashioned way. All of the finished boats were beautiful.

Steve

As the four of us walked by one of the waterfront restaurants, we saw other members of our flotilla inside enjoying a late-afternoon lunch. We joined them and ordered the house specialties of squid, lamb chops, veal scallops and eggplant.

The town of Spetses was a ten-minute walk around the harbor. The wind still blowing hard and quite cool, wearing jackets over our sweaters was a must. Many of the stores closed during the siesta period, but we window-shopped and found a couple of places to buy postcards and change U.S. dollars into Greek drachmas. The exchange rate was 350 drachmas for one dollar. One hundred U.S. dollars put 35,000 drachmas in your pockets and made you feel like a millionaire.

Back at the boat, the Starks napped, while the Baers explored more of the village sights. At 8:00 p.m. Dave returned to tell us he and Terry found a cozy café nearby and were enjoying a cup of cappuccino. We joined them and spent a snickering interlude talking about our adventures and our shipmates Steve and Marian.

Steve and Marian were from Seattle and almost defied description — well, almost. Steve was fifty-five years old, worked in the software industry, and fit the classic description of "computer nerd." He walked with his feet at a forty-five-degree angle to each other and slouched and shuffled like the Dirty Ol' Man of *Laugh-In* fame. Steve mumbled in a language indecipherable even to his wife, who constantly screeched in a chalk-on-blackboard voice, "Whaaat? Whaaat?"

Steve had a favorite shirt, worn for ten days straight, the tail of which was always out, sleeves unbuttoned, one rolled to the forearm, the other hanging down several inches below his fingers. His khaki pants hung on his Shmoo-shaped body, defying gravity. We held our breath in anticipation of the inevitable sighting of "butt crack."

At one time, Steve taught sociology at the college level and was once interviewed by Indiana University. Steve said, "IU was looking for someone a little unconventional, but they weren't ready for me! I didn't get the job."

Marian, not Steve's first wife, or perhaps his wife at all, was a West Coaster. Outspoken, liberated, Marian was not one to be trifled with. If ever cornered in a sleazy bar by a knife-wielding, drug-crazed thug, Marian is someone you would want watching your back.

Steve and Marian did not care for anything mainstream or ordinary — like guided tours or history lectures. Both preferred to do their own thing and went off on their own as soon as our boat was in port. Steve and Marian's presence was a memorable addition to our Greek experience.

We returned to the boat at 9:30 to find Wayne and Nan enjoying a quiet bottle of wine, and we joined them. After a short while I turned to journal writing, Dave to book reading, and Terry and Michele to card playing and giggling. It had been an adventurous day.

Our GPSC hosts declared that this would be a free day, the boats would remain in port. Although the morning dawned bright and clear — and remained that way all day in Spetses — surrounding areas, we learned later, experienced stormy conditions.

When we had come in, we noticed that the lighthouse on the point had some interesting statuary, and I walked over for a closer look. Numerous life-sized animals — bulls, goats, sheep, etc. — had been made out of welded steel. In addition to the animals were a mermaid with anchor-chain hair, and a statue of Bouboulina. Bouboulina, a female Greek warrior from this region, was legendary for her bravery and fighting prowess. She is referred to as the Greek Joan of Arc. Numerous streets and buildings are named in her honor.

The island of Spetses has a twenty-four-kilometer (about fifteen-mile) road encircling its perimeter. Michele and I planned to rent motor scooters and explore its circumference.

The scooter rental shop was very casual about its requirements. Helmets were nonexistent but after a test spin down the road — at our suggestion — we guessed we were ready. After we asked, we learned we would be responsible for gas. "Incidentally," the proprietor explained, "get it quickly, because the bikes are nearly empty."

Motorcycle Mama (Michele)

After exploring a few dead ends, we finally found a gas station, groped our way out of downtown, and started around the perimeter road. Michele led the way and handled her little scooter like a native. That's a compliment when you see how fearlessly the natives jockey their machines through traffic. I found myself struggling to keep my hat on my head and the wind-driven tears out of my eyes. Motorcycle Mama rides again.

Spetses is a rock-bound island with many hairpin curves and steep climbs and descents along its pine-shaded shores. Every few kilometers a beautiful turquoise cove would reveal itself just off the coastal roadway. We stopped at the gate of a palatial island villa to take pictures and catch our breath. Having shut off the machines, Michele's scooter refused to start when we tried to leave. I tried everything; it would just not make electrical contact. Two gardeners working on the villa property inside the gate responded to our request for assistance. They pushed the gate's remote control switch and came out to the road to help.

They were as perplexed by the problem as we were and pushed and twisted every lever and switch. At last, one of them inadvertently depressed the hand brake, and the scooter came to life and started. It was a safety feature; the brake had to be applied in order to start the machine. The two gardeners then gave us a ten-minute lesson in starting the motor scooter as if they had known it all along. They were not being smart, just being helpful.

After two hours of picture taking, beach admiring, and scene gazing, we completed the circuit and returned to the village from its opposite side. Once again, Michele charged through the narrow village streets like a Hell's Angel gone mad. She zipped around corners, scooted under the heads of buggy-drawing donkeys, and outmaneuvered local tradesmen on their motorcycles and pickup trucks. I thought to myself, She is either quite good at this or doesn't know how to apply the brakes.

Eight or ten of our fellow boaters were eating at a sidewalk café as we blasted by on our way across town. They cheered and shouted in answer to our beeping horns. The Baers had also rented a scooter, but their circumnavigation of the island had been out of sync with ours, and we did not meet. Steve and Marian also attempted to get scooters, but, as Marian explained later, she was somewhat fearful of the machines and opted to take a pedal bike instead.

Steve took the motorized version and discovered he had the heart of a motocross racer hidden behind his pocket protector. Later, when Steve rode up to our boat on his HOG, he had a grin on his face that stretched from Spetses to Athens.

Lynda and Kostas hosted a cocktail party on their boat at 5:00 p.m. and invited all GPSCers. Twenty-six people on the deck of a sixty-foot boat took footwork and balance.

Midway through the party, a thirty-one-foot sailboat approached our mooring, trying to figure out how to tie up. The visiting boat received a lot of advice from GPSC's twenty-six experts, their hospitality now enhanced by copious quantities of ouzo and wine, and was soon secured to our bow.

Michele discovered that the boat had five people from France aboard. Two were airline stewardesses with Air France. After a brief conversation with them, she returned to our boat to tell George, our captain, about the stewardesses. George was interested. When asked if he spoke French, he said, "I can learn." And within minutes, George had delivered a tray of wine and was on their boat deep into animated facial expressions and eloquent gestures.

Last seen that night as we turned into our cabins, was George, arm-in-arm with the two stewardesses, walking down the quay learning French.

This day we would make a quick transit to the port of Porto Cheli, only a short forty-five minutes away. There Adonis has arranged for a bus to take us on a day-long tour of the ancient city of Mycenae.

Oldest bridge in the world, built 1400 B.C.

The transit from Spetses to Porto Cheli was made under power, directly into the wind. Seas were running close to six feet and crashed over the bow as we plowed through them. Most of my shipmates were down below, keeping dry in the galley. I preferred to stay topside. I love these conditions and could almost hear *Victory at Sea* music as I wiped the salt spray out of my eyes.

We boarded the bus at 10:00 a.m. for a ride to the city of Mycenae. En route we passed what is reported to be the oldest existing bridge in the world, built in 1400 B.C. The bridge is quite small but still performs a utilitarian purpose, the means to cross a ditch.

The acropolis of Mycenae was rebuilt in the fourteenth century B.C. The ruins consist of partial stone walls, outlines of burial areas, and the remains of the palace structure. The entrance to the city, called the Lion Gate, is still nearly complete, and very impressive.

Adonis speaks of this history with such passion and enthusiasm that soon eavesdropping tourists start joining our entourage as we climb over the landmarks.

The mystical Cyclopes, one-eyed monsters of the Ulysses story, were fabled to have built these walls by lifting the huge boulders into place. Adonis offered an explanation of that fable. He suggested the Cyclopes were the engineers of the project and set their bearded faces apart from the bearded faces of the workers by a round tattoo in the center of their forehead—thus the one-eyed legend. As engineers they no doubt had methods (levers, winches, etc.) of moving the large boulders. Therefore, Cyclopes built the walls by lifting the boulders.

Adonis, an admitted agnostic, also had explanations for mysteries of the Bible that I found fascinating. Astronomy, about which ancient cultures studied most diligently, offered numerous coincidences to miracles found in the Christian Bible as well as similar events found in other religions.

Also located within a mile of the Mycenae ruins was the tomb of King Agamemnon. I find it difficult to separate fact from Greek mythology. Here in front of us were the stone and earth of a 3,200-year-old tomb. Yet, as real as the stones are, the story of Agamemnon is found only in Aeschylus and Homer.

Perhaps closer to the facts, Adonis suggested, the Beehive Tomb, thought to be the tomb of Agamemnon, wasn't a tomb at all, but rather a chamber used for religious celebration. Adonis said the entrance to the chamber had markings of a door that had been used numerous times, but a tomb would have been built to be entered only once.

Our next stop was the town of Nauplia, one of the most beautiful towns in the Peloponnese. It was Greece's first capital after the country regained its independence in 1827. The Palamidi fortress, an excellence specimen of the seventeenth-century Venetian fortifications, almost completely restored to its original form, dominates the town. Traditionally, 999 stone steps lead up to Palamidi. We would not make that hike this day.

The city of Nauplia is on a hill adjoining the fortress. From the apex of the city, we had a good view of both the fortress and the harbor below. It would be another wonderful place to escape with a stack of books and a sailboat.

There hadn't been much sailing on our cruise. Even though we lived on the boats for ten days, most of our transits were under power. Even on days of moderate wind, our schedule required us to get to the next stop at a specific time. The emphasis on this cruise was on touring and not, as I'd envisioned, sailing. That may be okay. Years from now, our memories will be of the once-in-a-lifetime sights we have seen, and not the sailing which we could do anytime, I suppose.

We left the Porto Cheli harbor at 9:30 a.m. on smooth water. Our destination, some forty nautical miles away, was the island of Poros. After motoring for an hour or so, the first stirrings of a breeze began. I was at the helm at the time and, with five knots of wind indicated on the anemometer, was able to coax 2.7 knots of speed out of the boat. Then slowly the breeze increased, first to seven knots, then to eight, then ten, and eventually to sixteen knots. All right! For the first time, the sailing had become exhilarating.

The other boats were widely scattered, some ahead and some behind. All enjoyed the fresh breeze, and all were thinking, Beat those other boats!

Up ahead I saw a point of land. It was apparent that if we kept clear of that point without tacking, we would become the lead boat. It was not a declared race, of course, but it was obvious our skipper wanted to be first. I worked hard, pinching as close to the wind as possible to clear the point. And it looked like we would make it.

Wayne, however, wanted a turn at the helm. Wayne, I observed earlier, was one of those helmsmen who thought that if he had the wheel in hand, he ought to be turning it. Unfortunately, when you do that, the rudder turns first one way, then the other, which has the effect of putting on the brakes.

Oh well, we're out here to have fun, and Wayne paid his money just like the rest of us. You've got the helm, Wayne.

Within minutes our speed was cut in half, and we needed to tack. What the heck, I had two hours of good sailing, and that was fun.

Poros is a lovely little island. The town wraps its arms around the harbor, resting its chin on the water's edge. Although it looks much like a classic Greek fishing village with boats, nets, and fish vendors, there was also a younger, yachty crowd there as well. Our four GPSC boats were a small percentage of the many gleaming charter sailboats and privately owned racing boats tied to the seawall. Most of the racing sailboats had sponsors' names prominently inscribed across the sails or along the length of their hulls. There was also a bevy of svelte, deeply tanned, young people walking along the waterfront as well. I liked the place.

Tied up alongside us was a seventy-five-foot motor-sailer with what appeared to be cabins and deck made of pine. Pine is not as durable as other wood you might expect to find in a maritime environment. We asked Captain George about that. His comment was, "It's a junk boat, you could buy it for $75,000."

As we were relaxing, a young British couple came aboard the pine boat. As we visited over the rail, we learned their boat was part of a new company, offering scuba diving charters. There were several partners involved. The couple was going to take the boat to Malaga, at the western end of the Mediterranean, where a client had signed on for a week. Malaga is a long way from Greece. The couple admitted their boat wasn't very seaworthy and that they would have to restrict their travel to fair-weather days only. They expected it would take three or four weeks to reach their destination.

The boat, they explained, was still under construction. We asked if we might come aboard for a tour. Although reluctant because of its unfinished condition, they saw us as potential future clients and welcomed us for a visit.

We stepped around the table saws and sawhorses and were amazed at the number of bunk rooms and the size of the internal spaces. It looked as though there were accommodations for thirty or forty passengers on that vessel. If we had been one of the voyagers, we agreed later, we would have stayed on deck close to the life raft. I noticed that for electrical wiring, they'd used Romex cable, an interior residential building wire, to connect all the outside mast and deck lights. Good luck with that.

Dave wished the young couple well and told them he was sure they would be successful. Later he said he felt sorry for such nice young people and was certain they had become linked to a guaranteed financial disaster.

Our departure for Athens on the last day would not take place until noon, giving us plenty of time for breakfast and last-minute shopping.

The transit would take five hours. I hoped for one last day of good winds, but as we left Poros, the sea was smooth. Dave and I sat on the forward deck, trading the last of our Navy stories. Michele and Terry talked on the galley roof behind us. We couldn't hear what they were saying but, from the occasional eruption of giggles, knew they were probably talking about us.

As we approached Athens, we saw a large group of sailboats, all sailing behind colorful spinnakers. They were competing in the Poros-to-Athens race held each weekend. We didn't have a spinnaker but, encouraged by the racers, raised our sails for one last attempt at sailing before concluding our cruise. The boat reached a respectable four knots, and we enjoyed the enlarging view of Athens as we approached.

Kalamaki Harbor in Athens is huge. Thousands of tall masts were seen peeking above the breakwater as we got closer. It was here that GPSC had its main fleet, as did Moorings and a number of other charter companies. Before unloading our gear, Veronica, the GPSC representative, boarded our boat and apologized for the problems we experienced early in our cruise. She presented each of the couples with an envelope and a gift-wrapped package. Later we discovered it was a credit for a 25 percent discount on any future charters with them. The package was a ten-pound ceramic dish. Just what we needed in our luggage for the trip home. It was a nice gesture, although the dish never left Greece.

Sailing among the Greek isles had been a wonderful experience. Doing so during the Easter celebrations added to the discovery of the rich Greek culture. We loved learning Greek history and seeing thousand-year-old artifacts. However, it is the memory of those hillsides surrounding Hydra, with its white villas, gardens, and walkways, all covered with bougainvillea, roses, and poppies, looking down on the turquoise waters of the Aegean Sea, that I will carry with me the longest.

# # #

Chapter Thirty-Five

Doing the Big Apple

Between 1978 and 1981, I ran seven marathons: the first in Huntsville, Alabama, to qualify for Boston; then Boston; followed by marathons in Chicago, Toledo, Terre Haute, Crawfordsville, and Detroit. During that same period, I submitted applications to get into the New York Marathon. It was a premier event. The twenty-six-mile course ran through the five boroughs of Staten Island, Brooklyn, Queens, the Bronx, and, for the finish, Manhattan's Central Park. The city closed some of its streets during the race. It was a big event and  rivaled, New Yorkers hoped, the prestigious Boston Marathon. Sixteen thousand athletes ran the race, four thousand of whom came from foreign countries just to run the Big Apple. Applicants achieved entrance by lottery. Sixty thousand runners applied every year. In 1982, after two failed attempts, I was selected. The experience was like I had indeed won the lottery.

Running was my avocation, second to my vocation as vice-president of our company. I was on its board, and the October 1982 date for the New York Marathon took place on a Sunday, one day before a board meeting scheduled the following Monday. It was mandatory that I attend the board meeting, and in casual conversation I mentioned the unfortunate timing of the New York opportunity to Jim Risk, our CEO.

"Not a problem," Jim replied. "Actually, there is a business in New York I'd like you to visit. Therefore, the company will fly you out Saturday, you can make the visit, run the marathon on Sunday, and fly back that night for the board meeting on Monday."

How about that for an accommodating boss!

Verrazano Narrows Bridge, NYC

When I arrived in NYC, I had the cab drive by the New York business Jim mentioned to satisfy the feigned business expense, on my way in from the airport.

The marathon began on Sunday morning on the Verrazano-Narrows Bridge connecting Staten Island to Brooklyn. Runners had been bused to Staten Island two hours prior to the race to be in place for the starting gun. We huddled in small groups under tents, trying to keep warm in the predawn chill.

Hundreds of porta pottys had been hauled in to accommodate the anxious participants. In the men's section, a mile-long trough was installed as a urinal. As I stood at the trough, it was humorous to see that previous NY runners, upstream, had launched matchbook-sized sailboats in the trough, their sails inscribed with encouraging messages: "Oh what a relief it is," "Aim to Please," "See you in Central Park."

I sat in a circle of six other runners waiting for the race to begin. There was a young girl from New Jersey who was running her first marathon because her whole family ran. In fact, both her mom and dad ran New York the previous year. Did really well, too, for old folks, she said. Finished holding hands. The crowd loved them. Oh, their ages? Both forty-eight.

Our group included a Spanish doctor from Brooklyn, a student from New Jersey, a fragile artist type from Paris, a young black man from Harlem, and a businessman from Australia, all folks I have great warm feelings for but would never see again.

All became curious about the small tape recorder I had Velcroed to my running shorts. A few marathons earlier, I started recording my runs, at first for the purpose of recording split times at the mile markers, but I found my comments during the race about what was going on about me and the sensations I experienced to be interesting in evaluating and reliving the experience. Some recordings were especially fascinating, like the runner's high I experienced (more like drunkenness) during the Toledo Marathon.

The Verrazano Bridge has two levels, six lanes wide. On one level the New York Road Runners Club, who organized the race, placed all the world-class and talented runners. Bill Rogers, who won Boston that year, was on that level. The also-rans, and that would be me, were on the second level. Using my Walter Cronkite voice, I announced into my recorder. "We're at the start of the 1982 New York City Marathon. It is a gray day, fifty degrees, with the hint of rain in the air. Perfect for running."

I was aware that those standing close by speculated on who this media reporter recording the event might be. Then the cannon fired, and we were off. After training hard and achieving some faster 10K and mini-marathon racing times, I was looking forward to setting a new personal record for the twenty-six-mile race. I wore a new pair of New Balance 100 running shoes. Most of the popular shoes — Brooks, Tiger, and Adidas — cost thirty or forty dollars. New Balance introduced a shoe with a price tag of one hundred dollars, the NB 100. I had to have a pair. Must be at least three times better, right? I knew not to put on new shoes without first breaking them in and ran several weeks in my 100s, but I now expected miraculous results with my new secret weapons.

What excitement! From the moment I left the bridge and entered Brooklyn, there was never a place along the race course that spectators weren't lined up five deep. *Chariots of Fire* and the theme from *Rocky* blared from every street corner. Local neighborhood fire departments shot their trucks' firehoses over the street intersections. At first the cooling water felt good, but then my feet became wet, and I could tell I started slipping in my shoes.

I'd visited New York many times while growing up in Connecticut but never realized how hilly it was, particularly when crossing the bridges from one borough to the next. The bridges had a steep initial ramp and an abrupt descent when you exited. My feet slid around in those hundred-dollar shoes, and blisters soon became open wounds.

At the sixteen-mile mark, after two hours and sixteen minutes, loudspeakers announced that Bill Rogers had won the marathon. Good for you, Bill. I had another hour and half to go. When I looked down at my wet shoes, I could see they were stained with blood.

My replay of the recording of the race later, from twelve miles on, revealed I was obviously hurting. My breathing was labored, and the mile splits were getting slower. After the announcement of Rogers' winning, I recorded, "My feet feel like hamburger. Ten more miles to go. Come on, you can do it. Uh-oh, it feels like my toe just came off."

The last few miles were painful. Finally, we made the turn along Central Park with only two miles to go. As I ran by the Plaza Hotel, I clearly heard someone shout, "Hey, Jim Stark!" I wondered if it might be my sister who lives in Connecticut, but it turned out it wasn't. I never learned the owner of that shout.

Central Park was incredible. All you could see was a sea of people lining the course. A recording of Frank Sinatra singing "New York, New York" filled the air. The joyful exhaustion brought many runners to tears. My time of three hours and forty-three minutes was twenty-five minutes slower than other marathons, but I finished. All runners were draped with a foil solar blanket as they finished and had large commemorative medallions hung around their necks.

After I made a brief stop at the refreshment area for Gatorade, a banana, and a health bar, I needed to get back to my hotel to shower and change into traveling clothes for my flight home.

I hadn't lost a toe, but my feet were raw from the wet shoes and abrasion. After taping them up, and thinking about the three-hour flight home, I rubbed myself down with analgesic balm in hopes the heat would relax my muscles while immobilized in the airliner seat.

I caught a cab outside the hotel. On the way to the airport, the cabbie grumbled about the traffic. "Christ, I can't believe it. They had half the streets closed today. Something must be going on."

"Yeah," I said. "The New York Marathon. I just ran it. See my medal?" It still hung around my neck.

"Oh, so that's it?"

"Yeah, and that's why I've got the analgesic smeared all over me. You must have smelled it."

"Oh, so that's what it is. I thought you had Chinese food back there."

Huh? I didn't ask for clarification of that one.

My tape recording of the New York Marathon has been listened to by a number of people. A friend, track coach at Columbus North High School,used to play the tape for his runners when teaching the "no pain, no gain" concept. I listen to it myself from time to time and think, "The thrill of victory and the agony of da feet!"

The following is part of an article I wrote for *Hoosier Runner*, an Indiana running newsletter:

"Running the Big Apple"

Doing the Big Apple, as in running the New York City Marathon, is a cornucopia of experiences, each of which is unique and everlastingly memorable.

New York Is Winning
The 16,000 chilled marathoners who lined up October 24, 1982, for the 26.2-mile tour of New York City, had already beat out 44,000 other applicants for those same starting slots. An early-postmarked entry and the luck of the draw had determined this dedicated jogger from Columbus, Indiana, was more qualified to run the streets of Brooklyn and Bronx than numerous other more talented runners.

New York Is Big
Logistically, New York is mind-boggling: twelve hundred medical personnel, 4,500 volunteers, 284 gallons of course marking paint, 68,000 safety pins, 500,000 paper cups, and the world's largest urinal. The race administrators process the huge number of runners by issuing bar codes to all official participants. Just like a package of wieners at the grocery, it's the scan of your bar code that triggers the instantaneous printing of your race number, the distribution of your information

packet, your check-in at the starting line, and the documentation of your finish at Central Park.

New York Is People and Camaraderie
It's 3,912 foreign runners getting their first exposure to the U.S. through our country's self-appointed goodwill ambassadors, the New York cab drivers.

It's two gorgeous snow bunnies from Denver, Colorado, who came East to run their third New York Marathon.
It's ten people pressed into the corner of the tent on Staten island spending the eternal two hours prior to the starting gun by opening their hearts and souls to the perfect strangers around them.

New York Is Excitement
When the cannon fires sending the army of waffle-soled combatants on their five-borough crusade of the city, the intensity and pitch of the spectators' encouragement is awesome. Music is heard blaring from every street corner. New York is on display. Flags, banners, and signs are everywhere. Bells ring, sirens wail, and people cheer like you can't believe.

New York Is Emotion
Along with the thrills, the gaiety, and the fun comes the inevitable marathon pain. Twelve hundred medical personnel aren't there for their health.

Some of my fellow athletes are reduced to a survival shuffle. Others show their badges of courage staining through their shirts and shoes, and still they continue.

At last the turn into Central Park, the final four hundred yards. Look at that crowd. Listen to that applause. Is that Frank Sinatra singing "New York, New York"? This is unbelievable.

Why, some of those runners are crying. Can you imagine that? No, not me. Must just be a cinder in my eye.

#

Chapter Thirty-Six

Magic Carpet

Eric and I were eager to get started on our two-week cross-country flight. Our Cessna 182RG clawed through the sodden clouds, occasionally bumped and jostled by minor turbulence on our climb to 8,000 feet. The weather briefing that morning had indicated cloud tops at 6,500 feet, with clearing conditions southeast of Indiana toward our first stop at Hot Springs, Arkansas. The rain lessened and the grey surrounding us lightened between layers as we passed through 5,000 feet. And then, there it was, just as forecast at 6,500, blue skies and glorious sunshine. No matter how many times I've had this experience, the breakout always brings a smile and a small sigh of relief to my lips.

One of my fantasies ever since I became a pilot was to fly cross-country in an open-cockpit, bi-wing airplane, landing in farmers' fields and sleeping under the wing. Those were the not-so-whimsical scenes for barnstorming pilots during the early days of aviation. In modern times, with FAA rules to follow and licensing requirements, just taking off into the wild blue is not so easy, but still doable if you know the regulations.

The FAA's Visual Flight Rules (VFR) allow cross-country air travel without flight plans as long as you adhere to proper altitudes, stay out of controlled airspace, and follow proper procedures when landing. Rural areas are abuzz with VFR pilots out for an afternoon of "flight-seeing" or heading off for their favorite airport restaurant for a $100 hamburger. That is an often-heard pilot expression, arrived at by adding the cost of fuel, although now the cost of that sandwich ought to be increased to $200.

In the summer of 1986, I was obligated to attend an industry convention in San Antonio, Texas. Eric, home from his first year at Wabash College, liked to fly; why not fly myself and Eric to Texas, fulfilling my business obligation, and then keep heading west, realizing my barnstorming fantasy? It wouldn't be in a bi-wing airplane, but my Cessna Skylane was a good second choice.

Cessna C-182 RG

Eric and I took the rear seats out of the plane and filled the space with bicycles, a tent, sleeping bags, and assorted camping gear. We just might spend some nights sleeping under the wing. Our itinerary included Hot Springs, San Antonio, El Paso, the Grand Canyon, Palm Springs, Santa Catalina Island, Death Valley, Salt Lake City, Ogallala, Nebraska, and home.

Commercial airline travel often takes more time to check in, go through security, wait for departure, and claim your baggage at the destination than the flight itself. Flying our own aircraft and having bicycles with us meant that after landing and ordering fuel for the next day, we could be on our way within minutes.

We selected Hot Springs for our first landing, as it was a midpoint on our way to San Antonio. We would be there only one night, so with an early departure planned for the following morning, we opted for a motel rather than camping.

Hot Springs is surrounded by the Ouachita Mountains, and we rode our bikes to a lofty peak looking down on the city. We were marveling at our bird's-eye view of the metropolis below when Eric said, "Isn't it interesting that even though flying in from eight thousand feet, we are now awestruck at being a thousand feet high on a mountaintop?"

Hot Springs, AK, 1986

    French settlers discovered Hot Springs in 1673; they found the healing properties of the thermal springs popular with Native Americans. Recent radiocarbon dating has established that the water reaching the surface began as rainfall 4,400 years earlier. Water soaking deep into the earth becomes superheated and rushes rapidly to the surface to emerge from forty-seven hot springs at a temperature of 143 degrees. The center of the town surrounding the magic waters was set aside as a federal reserve (later a national park) in 1820. The town developed into a successful spa in 1832, and in later years a number of art deco bathhouses were built to accommodate the tourists. The springs have drawn people to this area ever since.

    Perhaps even more famous than the hot springs was the illegal gambling that took place in the city from the late 1800s up until the 1960s. No fewer than ten major casinos and numerous smaller houses operated wide open. Hotels advertised the availability of prostitutes and off-track betting. Local law enforcement, including the Hot Springs police department and the Garland County sheriff's office, were paid to look the other way. Gangsters flocked to Hot Springs, and killings became commonplace during the 1920s. Gambling was finally closed down in 1967.

Eric and I rode the streets of Hot Springs, stopping occasionally to test the temperature of the water in the creeks.

"Stick your hand in that, creek, Eric. I dare you."

"Yeow! It's hot!"

The flight to San Antonio went without a hitch. Once we were there, the four-day convention kept me busy with business meetings and cocktail parties. Eric and his mother, who had flown in commercially to meet us, toured the San Antonio attractions, so as to remember the Alamo, the River Walk, and other touristy sights.

Finally, the real adventure began.

Saying "Adios" to San Antonio, Eric and I soon winged it west again at minimum altitudes, enjoying the scenes of desert flora, fauna, and open-range ranches.

Eric asked, "Have you always loved flying, Dad?"

"Not really. Prior to the Navy, I never even considered flying. So I guess it all started with that experience."

"What do you like about it?"

"Hmmm, I've wondered that too. Is there a common appeal about airplanes, motorcycles, and sailboats, other than wind in my hair? Partially, I think, it's the pride in the skill to operate such conveyances, but then there's that feeling of freedom. How marvelous it is to be zipping along above these desert plains, with all points of the compass at our wish or whim. I have that same free-spirit feeling when traveling on my motorcycle or sailing off toward shoreless horizons. How cool it is."

Next stop, El Paso on the Mexican border. I had considered filing a flight plan to enter Mexico, but procedures for entering Mexican airspace were thorny, and should there be hassles, I didn't want my airplane stuck on the wrong side of the Rio Grande. We did, however, walk across the border into Juarez.

It's been my experience on at least three other occasions that Mexican border towns are generally not attractive places. Unlike Guadalajara or Mexico City, which have beautiful architecture and landscaping, the border towns seem to have shabby neighborhoods, noisy honky-tonks, and ornery-looking hombres. After a quick meal of genuine tostados, we returned to the more familiar and comfortable USA.

Our next stop was Grand Canyon National Park. Flying up from the southeast across New Mexico and Arizona had us traveling over vast areas of unpopulated open prairie. Visual Flight Rules required an airplane to remain one thousand feet above congested areas, but open areas such as these allowed us to drop down to five hundred feet. That's five hundred feet above terrain, so pilots have to remain aware of rising hills, cell towers, and power lines. But zipping along close to the ground gives you an exhilarating sense of speed and a close-up view of cattle, cactus, and earthy wonders. In the Navy we called it flat-hatting whenever you dropped down to treetop height. I remember one of my fellow student pilots who got into trouble for flat-hatting after one of his solo flights and wondered who had seen him and turned him in. Later he found out it was the leaves and twigs sticking out of the leading edge of his wings that gave him away.

We didn't fly that low, but I admit to a couple of drop-downs for a closer look at a raft on a river in one instance and to wave to a couple of cowgirls in another.

Grand Canyon National Park Airport is the only state-owned airport in Arizona. As we approached from the south, we could see the edge of the canyon just north of the airstrip, but rather than do a fly-over, we chose to land, load up our bikes with camping gear, and ride to the campground before all sites had been taken for our two-night stay.

The Canyon Road closely parallels the canyon. At the first opportunity, we pulled off to get our first look over the edge. What a sight! It took my breath away. The enormity of the depth and width is staggering. The Grand Canyon is huge, or, dare I say, grand! It's measured as 277 miles long, as much as six thousand feet deep, and ten miles across in places. But it's the rich variety of red, orange, and gold colors that provokes your "Wow." There must be a more descriptive exclamation than "Wow," but both of us stood there open-mouthed, expressing that very word. Just remarkable.

The Grand Canyon

Climbing back on our bikes, we didn't ride more than five minutes before we had to take another look. "Wow!" We barely made it to the campground, needing to stop every few minutes to take in that wondrous sight.

That night at our campfire, under a canopy of stars more brilliant and plentiful than ever before experienced, Eric said, "For the rest of this trip, if we don't see another sight a tenth as beautiful as the Grand Canyon, just being here today makes it all worthwhile."

"Amen to that."

The next morning, after a breakfast of freeze-dried scrambled eggs — not bad actually — we mounted up to head for the visitor center to plan our day. Among the many trails available to hike down into the canyon, Bright Angel was suggested as one we might enjoy.

Hikers are warned not to attempt to go to the bottom and back in one day. The top of the rim at Bright Angel is 6,850 feet. It's over three thousand feet to reach the bottom. Water is available at rest stations along the way, but hikers are strongly encouraged to carry their own, go slowly, rest often, and stay cool.

Most day hikers hike down a certain distance, then return. It takes twice as long to climb up as to go down. We planned to descend for an hour, then take two hours to get back.

The trail itself is narrow, three to four feet in width. Not only would people be passing us, but we would also pass mule trains. There are rules when a mule train passes:

- Step off the trail to the uphill side away from the edge.
- Follow the directions of the wrangler. Remain quiet and stand perfectly still.
- Do not return to the trail until the last mule is fifty feet past your position.

Hikers have been injured and mules killed on the trail. Over 250 hikers are rescued from the canyon each year.

Going down is easy, and there were ten or fifteen people spread out in front of and behind us. At the rest stops we became acquainted. There was Jack, a middle-ager from England; Tony and his new bride, Winnie, from Ohio; Mrs. Stewart and her two young daughters, Laura and Emily, probably six and eight, each wearing open-toed sandals; Pops, overweight, overdressed, and not wanting to be there except at the insistence of his grandson, who darted back and forth, brushing against people as he passed. Surprisingly, all of these folks moved at a fairly common pace while going down.

By the time Eric and I reached our agreed-upon turnaround point, almost all of our companions had started their return climb. Quite a difference going uphill. As we methodically placed one foot in front of the other, we began catching up and reeling in the others. Jack had turned when we did and matched our pace step for step. Mrs. Stewart sat on a rock, one of the daughters in her lap, crying. Pops didn't look so good. His brown, long-sleeved work shirt was drenched. "You okay, Pops?"

"Yeah, I'm fine, but when I catch that grandson of mine, I'm going to kill him."

There were rangers stationed along the trail, and we mentioned to one in passing that Pops about two hundred yards behind might warrant some watching.

The view was magnificent. I snapped picture after picture in hopes that my two-dimensional photos captured at least a smidgeon of the splendor. At last we made it to the top.

We spent the rest of our day pedaling the South Rim road and marveling at the vistas.

Upon departure the next day, I decided to take advantage of a once-in-a-lifetime opportunity and fly into the canyon below the rim. Indeed, it turned out to be a timely opportunity, because shortly thereafter flights over the Grand Canyon were prohibited. There had been aircraft accidents over the years. In 1956 a United DC-7 and a TWA Super Constellation collided over the canyon, killing 128 people. It was speculated that the pilots were treating passengers to views of the canyon. In 1986 a twin-engine plane and a helicopter carrying vacationers on sightseeing tours collided, killing twenty-five people and leaving charred wreckage on the canyon wall. Both aircraft were below the rim at the time of the collision.

Our flight occurred just three months after the helicopter and tour plane had crashed, and a short time before the FAA designated the air above the Grand Canyon, up to 18,000 feet, restricted airspace.

Flying over the rim and reducing the throttle to descend into the abyss was a physically and mentally unpleasant feeling. My stomach was doing flip-flops. Would the airplane respond when I applied power to level off? Why does the engine sound so rough? Never noticed that before! This is no place for carburetor ice or the other hundred things that could cause an engine failure. An emergency landing here was not possible.

We were no more than two hundred feet below the rim, but so was a tourist helicopter I could see up ahead. And what about other aircraft behind us, or below? I was not having a fun experience. Eric, on the other hand, thought it was cool. "Look at the mule train, Dad. They're waving at us. Can you go lower?"

"No!"

I've made landings with an engine out in stormy weather, flown through violent thunderstorms and icing conditions, and this was equally stressful.

"Okay, Eric. Up we go. We're out of here and on to Palm Springs."

After landing in Palm Springs, Eric and I popped out of the cockpit, expecting to inhale dry desert air, and were practically slammed back into our seats by the humidity. I asked the lineman, "Where's that dry climate I always hear about?

Palm Springs, CA, 1986

"It's those damn golf courses," he said. "Irrigation systems. Hell, we've got more golf courses here than traffic lights. Every week you read about a new one being built. There are over one hundred."

Palm Springs, just over a hundred miles east of Los Angeles and northeast of San Diego, is a resort town. It used to have a dry climate. Starting in the early 1900s, it attracted visitors needing dry heat for health reasons. Its popularity continued over the years, attracting successful tycoons, sports figures, and movie stars. Bob Hope, Frank Sinatra, Arnold Palmer, Joe DiMaggio, and numerous others had built homes in Palm Springs neighborhoods.

There were no campgrounds in the city, so Eric and I dragged out our bicycles and backpacks and headed off to find a cheap motel and tour the city.

Every street corner had an explosion of colorful flowers. The boulevards were lined on both sides with fifty-foot royal palm trees. Landscapers dressed in clean white coveralls were manicuring the lawns and gardens of the lovely homes and estates we were passing. We started counting Rolls Royces, Bentleys, and other luxury cars, but they soon became commonplace. "Ho hum, there goes another one."

We found a Howard Johnson's motel and paid twice what our other accommodations had cost. After dropping off our bags, we continued touring. We planned an early departure the next day for a flight over Los Angeles on our way to Santa Catalina Island.

Los Angeles and southern California constitute one of the busiest air traffic areas in the country. I could have filed a flight plan to fly through the area and be in direct contact with air traffic controllers. However, there are several visual flight corridors through the area, and since Catalina is only twenty miles off the coast, I opted to depend on my own eyes in the sky. Bad decision.

I had heard about the Los Angeles smog but didn't give it much thought when taking off into the blue skies over Palm Springs. We soon found ourselves flying through the clichéd pea soup, straining to see the airliners I listened to over my headset. This was not good, and I immediately began a climb in an attempt to find clear air.

My Skylane has a service ceiling of 18,000 feet. However, once you go above 10,000 feet, oxygen becomes a requirement, and we weren't carrying any. Passing through 9,000, we were still in the crud. Ten thousand feet was no better. It wasn't until punching through 11,500 feet that we were finally above the smog. Now, two FAA rules come into play: the first is that all VFR traffic heading westerly is to fly at even altitudes plus 500 feet. Okay, 12,500, no problem, I could do that, but the second rule states that pilots flying between 10,000 and 12,000 for more than thirty minutes must be on oxygen, and above 12,000 a pilot must be on oxygen continuously.

We were less than ten minutes from the coast, where I planned a rapid descent. "Keep an eye on me, Eric. If I pass out, let's hope you remember those flying lessons I gave you as an eight-year-old."

In truth, I wasn't too concerned. Oxygen-chamber training in the Navy demonstrated short-term deprivation wouldn't be a problem at that altitude.

Our ten-minute flight over LA was interesting as we watched airliners punch through the smog all about us on their ascents to 30,000 feet. Our climb above the soup had been a wise decision.

Once over the coast, the smog dissipated, but the sky was full of puffy white clouds with scarce blue space between them. I had tuned in the Santa Catalina airport radio frequency and could clearly hear the landing traffic. I had to find a hole to get down.

Finding one, I lowered the landing gear and flaps to control the airspeed and nosed over into a steep dive. Catalina lay just ahead, perhaps five miles away. Here's where I have to admit pilot error. I hadn't looked carefully at the airport chart. I assumed the airport would be at sea level. When I brought the plane to level flight at one thousand feet, we found ourselves looking at the steep rock sides of Santa Catalina Island. A quick glance at my chart revealed the airport, called "Airport in the Sky," was located on Catalina's mountaintop. "Oops. Okay, Eric, where is that hole again? We've got to go back up."

Seeing the airport and getting into the landing pattern was not difficult. The single runway, however, was a daunting sight. Renters of aircraft will not let customers fly out to Catalina without first making the flight with a flight instructor. The runway is only three thousand feet long, rather short by most runway standards. But its most interesting feature is that its approach end and far end begin and end at the edge of a cliff. Downdrafts are constant as you approach the runway threshold.

It reminded me of a time back in my Navy days when a three-plane detachment was sent to France to work a joint exercise with the French Air Force. They had a runway that began at the edge of a cliff. When we arrived, the entire French squadron came out on the tarmac to watch our landings. They knew about their downdrafts, but our pilots did not. The French guys took smug pleasure in watching our planes slam into the runway and bounce high in the air.

I anticipated the downdraft at Catalina but was more concerned about the short runway. My astute copilot had an excellent suggestion. "Dad, a landing not too short and not too long would be just fine."

Santa Catalina Airport

The population of Santa Catalina Island, often just called Catalina Island or simply Catalina, is less than four thousand, 90 percent of whom reside in the island's only incorporated city, Avalon. The city is nearly two thousand feet below the mountaintop airport. What better way to arrive than to coast downhill on our bicycles? After securing the plane and ordering fuel for next day's departure, we mounted up and headed downhill.

The road down felt as steep as the descent I made after flying over Los Angeles. The road has a series of switchbacks, hairpin curves, each having a convex mirror at its apex so you can see oncoming traffic. Our brakes were smoking but thankfully holding, because, coming out of the first turn, we found a dozen buffalo standing in the middle of the road. Buffalo on this California island? What's that all about?

We found out later that buffalo were not native to Catalina but first imported there in 1924 for the Zane Grey silent movie *The Vanishing American.* The buffalo grew into a herd of six hundred and were periodically relocated to the Great Plains. It was determined that 150 was the ideal-size herd and is now maintained with birth control.

The chewing-gum magnate William Wrigley, Jr., bought the island in 1919 and invested millions of dollars to turn it into a tourist destination. He built a casino and established steamship service to transport guests to Catalina. Wrigley also owned the Chicago Cubs baseball team, which used the island for their spring training. In 1975 the Wrigley family turned over 90 percent of the island to the Catalina Island Conservancy, which they had established three years earlier, but they maintained control of the resort properties.

After our exhilarating ride, Eric and I checked into a bed and breakfast and walked the town, enjoying its sights.

The casino on the water's edge was built in 1929 as a dance hall and movie theater. Its $2 million expense included Tiffany chandeliers, huge art deco–style murals of stylized seascapes, and a 20,000-square-foot ballroom. My only regret on this western flight has been our all-too-brief stops at each of the places we visited. Avalon, despite its small size, was a city where you could spend a busy week. Its many beaches and coves offer a variety of aquatic activities, such as kayaking, sailing, and snorkeling. Fishing is also a popular attraction. The scenic island encourages hiking and biking.

I found my love in Avalon beside the bay . . .
I think I'll travel on to Avalon.

Eric and I had a delicious seafood dinner at one of the waterfront cafés before turning in for the night — not, however, before making arrangements for a cab to pick us up in the morning for transportation to the airport. We discussed trying to ride up the airport road on our bikes.

"I don't care, do you want to?"

"I don't care, if you want to."

"Well, if you don't care, I don't care."

"Okay, let's forget it."

"Whew!"

The next day's flight would include a landing in Death Valley.

I was looking forward to landing at Death Valley, a little over an hour away, because I had never seen my altimeter go below zero before. Our plane would be touching down at the second-lowest point in the Western Hemisphere, 282 feet below sea level. Someplace in Argentina is the lowest.

The valley is actually called a graben, sinking land between two faults. It's a huge area, over five thousand square miles. Located beside the Mojave Desert, it's the hottest and driest place in North America. With less than two inches of rain per year, temperatures soar in the summer months, averaging 120 degrees during July. A record 134°F was recorded in 1913. We arrived in August, and opening the door of the plane felt like opening the door of an oven.

Death Valley got its name in 1849 when a wagon train of California gold-seekers became lost trying to find a shortcut off the Old Spanish Trail. Unable to find a pass out of the valley, they were forced to eat several of their oxen and use the wood of their wagons to cook the meat. Eventually, survivors were able to hike out of the valley. As they left, one of the women turned and said, "Good-bye, Death Valley."

I mentioned Eric's and my arrival in this place in an earlier chapter. We planned only a brief visit before heading on to Salt Lake City, Utah. The Death Valley visitor's center and museum were located only a short walk from the airport.

Situated in an irrigated area of green grass and trees, the museum was also next to a campground. That small oasis was attractive, but I wondered what campers did for fun in the surrounding desert. Sunbathing would take only a few minutes. What about the rest of the day?

The museum was interesting. We learned about the several species of plants and animals that have adapted to the harsh desert environment. Some examples include creosote bush, bighorn sheep, and coyote. It was particularly fascinating to read about the moving rocks.

Mysterious moving rocks

The dry lake beds found throughout Death Valley are called playas. Racetrack Playa is known for its moving rocks, some weighing hundreds of pounds. The rocks, some buried several inches in the lake bed, have left trails of their movement hundreds of feet long, revealing the turns the rocks made. There are no footprints around the rocks to indicate human or animal involvement. The mystery remained unsolved for years but was finally resolved. It was determined, and actually caught on video, that during the winter months, on rare occasions, rain in the mountains would flood the flat playas in Death Valley with an inch or two of water. The playas' muddy surface would become slippery as grease. At night, when the temperature dropped and the winds blew, a thin skin of ice formed. The wind drove the inch-deep water and ice across the playas, carrying the rocks with it, leaving the telltale grooves in the surface. Thus, the mystery was solved.

Salt Lake City is surrounded by the Great Salt Lake to its northwest, and the high — 12,000-foot — Wasatch and Oquirrh mountain ranges to the southwest and east. Flying between the peaks, and still above 10,000 feet, we approached the Salt Lake airport. I was in contact with the tower as we approached, and the tower operator knew our altitude but wasn't giving us descent or landing instructions. With the runway only five miles ahead, the tower asked if I was able to make a straight-in landing (no traffic pattern).

"Affirmative," I replied.

"Roger, Seven-three six, you're cleared for a straight-in to runway niner. Please expedite."

It would be a challenge! I was still at 10,000 feet, and he wanted me to land on a runway practically under our nose. "Hold on, Eric. This is going to be a dive-bombing plunge."

Dropping wheels and flaps, we made our near-vertical dive, landing on the second half of the accommodatingly long runway. Welcome to Salt Lake.

Mormon Tabernacle and Temple, Salt Lake City

The city is beautiful. Brigham Young founded and designed it in 1847 as headquarters for the Mormon religion, Church of Latter-Day Saints. Young selected this place because it was beyond the jurisdiction of the United States. The Mormons had considerable turmoil in their previous locations in Kirtland, Ohio, and Nauvoo, Illinois, particularly in regard to their practice of polygamy. The Mormon founder, Joseph Smith, was killed by a mob after he was jailed, partly because he had tried to acquire a wife without the benefit of her earlier divorce.

Eric and I found accommodations for the night, then biked about the city, enjoying its clean, uncluttered ambiance.

The streets are wide, at the direction of Brigham Young, who wanted them wide enough that a wagon team could turn around without the driver resorting to profanity. Temple Square is the iconic symbol of Salt Lake, containing the Mormon Temple, built between 1864 and 1867. The Tabernacle is where the famed Mormon Choir performs.

The Tabernacle, separate from the Temple, was originally built to hold Mormon conferences and has a seating capacity of seven thousand. The dome-shaped building, constructed using a lattice-truss system, was held together by dowels and wedges. The dome roof, supported by forty-four sandstone piers, is 150 feet wide by 250 feet long. Skeptics predicted that when the interior scaffolding was removed, the whole roof would collapse. The roof is nine feet thick and has remained standing for over 150 years. The Tabernacle organ, the largest in the world, has 11,623 pipes.

Before the advent of electronics and amplifying systems, Brigham Young chose the elliptical design of the building so that sound from the pulpit would be concentrated and projected to the opposite end of the building. That was demonstrated by a coin dropped onto the wooden floor at one end of the Tabernacle being clearly heard 250 feet away.

Eric and I were then treated to the magnificent Mormon Tabernacle Choir. The 360-member, all-volunteer choir has been in existence since the Tabernacle was built, 150 years earlier. Members, many being husbands and wives, are allowed to belong for twenty years or until they reach sixty years of age, to give new members a chance to join. The choir is famous for its many albums, concerts, and world travels. It has performed for ten presidents, sung at the inauguration of six presidents, and won two Peabody Awards, two Freedom Foundation awards twice, three Emmys, and numerous other honors and recognitions.

I had heard the choir on recordings and radio and television broadcasts a number of times, but to be in that hall with the organ playing and those 360 voices singing is an uplifting experience. It moved both Eric and me with amazement and admiration.

"Dad, remember what I said about Grand Canyon? This may be even more inspiring."

I can't help but think the experience had a lot to do with my son becoming the conductor of the Indianapolis Symphonic Choir. He has since conducted at Carnegie Hall, Kennedy Center, and other venues around the world.

Salt Lake was definitely one of the highlights of our adventurous trip, but we were now on the homebound leg of our journey. Weather up till then had been amazingly cooperative. That Stark good fortune continued to shine upon us. Other than the smog over Los Angeles, weather was, as we liked to say in the Navy "CAVU," clear and visibility unlimited.

Leaving Utah, we had our first forecast of a developing low-pressure system with accompanying thunderstorms. A fast-moving front from the southwestern part of the country was rolling in; therefore, we set a northeastern course toward Ogallala, Nebraska, which had a campground on a large lake that looked inviting.

I had filed an instrument flight plan out of Salt Lake and was in constant communication with FAA controllers. The advantage of flying IFR (Instrument Flight Rules) was having continuous weather updates. Looking behind us, we could see dark skies. Up ahead it was clear blue with puffy white clouds.

Eric and the controllers laughed at my tongue-tied pronunciation of Ogallala. It was not until landing and practicing that word that I finally got the hang of it. Then it became a fun word to say, "Oh-gah-la-la."

After landing and tying double tie-downs in view of the approaching storm — we loaded our bikes and backpacks with camping gear and started out. Lake McConaughy is the state's largest lake and the largest reservoir in a three-state region. The recreation area's campground was north of the airport, twelve miles away. The black clouds moved in fast, and it was going to be a race to see if we could get to the campsite before the storm caught us.

Thunderstorms in Nebraska can be extremely violent. Loaded down as we were, racing up the highway was not easy. The clouds gained on us, their color changing from dark gray to black with a mixture of purple. The storm was rolling across the plains like an oncoming freight train, and the winds first blew sideways and then head-on, almost bringing us to a stop.

"Come on, Eric. It's just a couple more miles!"

"Just trying to keep from outrunning you, old man."

We entered the campground and headed for the welcome center. On arrival we dumped the bikes as the first raindrops, big as marbles, splashed against us. Not taking time to unload the bikes, we entered the building just as the skies opened up. And what a thrashing deluge it was! Trees bent over like limp noodles. Everything not tied down outside swirled about in the wind. The bikes, heavily loaded, remained grounded.

The storm lasted about 45 minutes. After registering for our campsite, Eric and I set up our wet tent, soon to be dry in the now-sunny skies and warm breezes.

How incredibly lucky we had been on this trip — no mechanical problems, no weather issues other than this near miss, and now, with only one more overnight stop, somewhere in Illinois, we would soon be home. What sights we have seen and fun we had. The tachometer registered thirty-eight flight hours. We had been gone fourteen days and covered over five thousand miles. My barnstorming fantasy had at last been fulfilled. It had been, indeed, a magic carpet ride. And how great to have shared this experience with my son.

# # #

Chapter Thirty-Seven

Wyoming National Parks

The park ranger's note on my tent was serious. I had violated a strict campground rule and received a stern warning. I left a container of water on my picnic table. Bears are a significant concern in the campground, and therefore all foods, food wrappers, and any product emitting a scent must be secured in the steel "bear box" provided at each campsite.

Oops, sorry about that. Didn't know Fiji Water was such an attraction.

The campsite was deep in the pines, yet only a short distance to the restrooms and shower facilities. The campground, located a few miles from the entrances to Grand Teton National Park and Yellowstone, would be an ideal location for the next five days.

Grand Teton National Park is unbelievable! The various lakes reflect the grandeur of the surrounding mountains. That first day I hiked a lakeside trail, taking pictures through the pines of the lake with its mountain backdrop. I fantasized I was taking photographs destined for *National Geographic* magazine.

I spent a cozy night bundled in my down sleeping bag with only my nose exposed to the brisk forty-degree temperatures. Damn, I love being out there in a tent with campfire, starlit skies, and the feeling of roughing it.

First on the agenda for the next morning was a ride to Jenny Lake Trailhead, where I'd take a boat across the lake to hike up to Hidden Falls. I could have hiked around the lake, but that takes an extra two hours. Although only a half-mile from the boat landing, Hidden Falls requires a five-hundred-foot climb. Already at seven thousand feet, oxygen for heart and lungs becomes an unfamiliar necessity. Our boatload of hikers started up the nature trail, led by a rather heavy-set lady who I knew was walking much too fast. I found her ten minutes later, sitting on a log, gasping for breath. She had gone as far as she would go that day.

The snows melting off Teton's upper reaches come rushing down in a flurry of whitewater. Hidden Falls drops one hundred feet from a lofty apex to tumble into creeks below. The roaring water rushing past us on the hike up the hillside made conversation impossible.

It's quite a sight and sound. Once reaching the base of the falls, all hikers take turns exchanging cameras for pictures of each other standing before the whitewater cascade.

The next day I rode into Yellowstone. Although only thirty miles from the campground to the Yellowstone entrance, it would be another sixty miles before I reached the first visitor center inside the park. This is big country; it's hard to get used to that.

The female twenty-something ranger at the visitor center pointed out all the scenic attractions I must see if I would be in Yellowstone for only three days. Her suggestions included several eight-mile hikes that I immediately put into the category of "in your dreams, sweetheart; you're talking to an old guy here."

My plan was to visit the geyser areas in the park the first day, then try some shorter hikes on day two. On the third day, I'd do a riding exploration, stopping at various overlooks, putting my photo portfolio together for the *National Geographic* submission.

First stop in the geyser areas had to be the world-renowned "Old Faithful." I had seen this wonder once before as an eight-year-old. It had changed little. The part I didn't remember was the throng that gathered for this every-fifty-minute (approximately) spectacle. With three million people visiting Yellowstone every year, I guessed the attendance to be 5,000 to 10,000 for the hourly show. All raise their smart phones and digital cameras to record the scene. I can't help wondering about the number of lost film sales these technologies have caused George Eastman's employees. Spectators applaud appreciatively when Old Faithful spews its final blast. I stayed for two performances, walking the surrounding pools in between eruptions.

Old Faithful, Yellowstone National Park, 2011

Old Faithful is located in what is called the Upper Geyser Basin. There are two others, the Midway Basin and the Lower Basin. Each has its own geysers and hot springs, fumaroles, and mud pots. The origin of all this eruptive activity is volcanism. Two million years ago, huge volcanic eruptions occurred here and continue today. Deep earth heat powers the activity. The hot springs and fumaroles are boiling hot. Signs warn tourists not to leave the safety of the boardwalks. The boardwalk around Old Faithful was funded by a family whose young son ventured off the designated walkway and was scalded to death. Its being a windy day, I saw that several ball caps had been blown into the area surrounding the geysers. They were not retrieved. Apparently people do read warning signs.

One of the remote roads off the main highway, called Fire Hole Lake Drive, takes you within a few feet of bubbling pools, right at the edge of the road. How hot is that water, anyway? I mused. Coming to a stop, I stuck my fingers into the pool. It reminded me of when I was a kid and my mother said don't touch, it's hot, and I touched it anyway.

Back during the turn of the nineteenth century, fly fishermen would catch fish in the streams, then turn around with the fish still on the line and drop them in the boiling fumaroles to cook. The government outlawed this practice in 1911, insisting the fish must be first killed with a club or a knife before boiling them.

Grand Prismatic Spring, Yellowstone National Park

The Midway Geyser Basin contains the Grand Prismatic Spring, the largest of all the hot springs. Its deep blue waters invite you to step in for a good soak. But the steam coming off the water, accented with the smell of sulfur, wafts across you, almost taking your breath away with its searing heat. Guess I'll stick to a warm shower.

I visited with my campsite neighbors, a grandfather and grandson from Pennsylvania, traveling on a motorcycle together, I asked the grandfather, a frequent visitor to Yellowstone, where his favorite areas were. He said he loved the northern parts with their wildlife and canyons. Yellowstone is huge. My park map suggested that if I was going to get to the places he mentioned, I was going to have to ride a hundred miles. Therefore, I revised my plans to allow plenty of time to explore the northern part for picture taking, short hikes, and vista viewing.

There are numerous spectacular waterfalls in Yellowstone, because of the volcanic activity. Cascading rivers from the mountaintops carve out deep holes in the earth and rocks, softened by the heated eruptions. The Lower Falls and Upper Falls drop three hundred feet into holes carved another one hundred feet deep. *Dazzling, just dazzling!*

One of my goals in Yellowstone was to see where the buffalo roam and the deer and the antelope play. My campsite neighbor said this was the area they gather, and he was correct. You don't have to look hard. The road up ahead would be suddenly clogged with dozens of cars, all their passengers hanging out of windows or over a guard rail, pointing their Instamatics and cell phones at deer, elk, and buffalo. I'm sorry to say I did not see one bear.

What a lovely thing it is to ride along beside the sparkling streams and see fly fisherman, thigh-deep in their pursuit. I considered bringing my backpacking fly rod but decided I had to draw the line somewhere. If that, then I'd want the kitchen sink. Come to think of it, I have one of those with me.

I spent most of the day at 7,000-to-8,000-foot elevations. Michele reported Indiana was having a heat wave, and here I was debating whether it was warm enough to remove my motorcycle jacket. Anytime a sign indicated I was above 8,300 feet, I could see patches of snow in the woods.

The sun is intense at this elevation, and my arms from the elbows down, as well as my cheeks, had become a leathery, deep copper color. Every morning I rubbed down with skin cream and SPF 30. My fingernails were all broken and filthy. I had a ragged scab down one leg after an encounter with a cactus, and my clothes became filthy after just one day of sweat, wind, dust, and bugs. This would not be a trip for someone who takes ninety minutes working on her hair to get ready for church. (Who could that be?)

I broke camp at Colter Bay and exited Yellowstone at its northern entrance. That took me through a small piece of Montana, then into Idaho. Yellowstone had been spectacular. (How many times have I used that word?) What a wonderful treasure it is for America. Yellowstone ought to be a required field trip for every American youngster. My three days there and my days in Grand Teton were just a quick glimpse of these fascinating places. Those staying longer could avail themselves of whitewater rafting trips, horseback rides, hiking the trails, lectures, movies, and much more.

As I entered Aston, Idaho, a very helpful visitor center guide suggested the attractions near Twin Falls were just what I was looking for: Ice Caves, Thousand Springs, and the Twin Falls themselves. "But first," she said, "you simply must not miss Craters of the Moon National Monument."

Fifteen thousand years ago, volcanic eruptions deep within the earth thrust lava across major parts of Idaho. The most recent eruption was two thousand years ago. The result was a landscape that appears similar to the surface of the moon. Calvin Coolidge designated this area as a national monument in 1924.

The visitor-center person said I could either take the main highway, U.S. 20, or I could ride eighty-five miles across Idaho's desert. It's very *interesting*, she said. Bingo! Lady, you just said the magic word.

Idaho's desert was not like you envision an Arabian landscape. It is covered with low brown grass, spotted with small shrubs. It is very flat, with mountains rising in the distance, mountains that seem to take forever to come closer.

Most noticeable of all was a southern wind that blasted me like a blowtorch the entire way. It was blowing at a steady thirty knots, and even though I was traveling eighty mph, I had to lean the motorcycle twenty degrees or more to keep from being blown off the road. My bike's outside temperature gauge showed only eighty-three degrees, but it felt searing. I stopped twice to smear sunscreen on my cheeks to ease the burning. Those mountains eighty-five miles away, my destination, simply wouldn't get any closer.

I finally completed the desert run, blasted, burnished, and bushed. My goal was Arco, a little town close to the Craters Monument, and to find a motel for the night. After four days of camping, I needed the comforts of an in-room shower and conveniences. The signs welcoming me to Arco boasted that it was the first town in the world to have its electrical services provided by nuclear energy. Hmmm, that's *interesting*. I asked the motel owner about it. He said they built an experimental nuclear reactor here in Arco and used it to power four electric lightbulbs.

"But the sign said they powered the city by nuclear energy," I challenged.

"Ah, yes," he said. "They then built a larger reactor, connected it to our power grid, and powered the entire city for four hours. They then shut it down, carted it off someplace, and now we're back to using coal."

"I see. So, where did the town get the name Arco?" I asked, expecting some kind of an acronym.

"People ask that a lot. The town debated a name for our village for some time, wanting something to do with *crossroads*. Then someone noticed an English duke by that name was visiting New York, so they used his name."

"Did the duke have anything to do with this town, or did he ever visit here?" I asked.

"Nope."

Chapter Thirty-Eight

Pony Express Trail

Prior to every lunar landing, astronauts were sent to Arco, Idaho, to train in a setting nearly identical to the one they will encounter on the moon. The area is known as Craters of the Moon National Monument

The landscape is remarkable. Lava flows from volcanic eruptions two thousand years ago covered 750,000 acres of Idaho. I don't know what that is in square miles, but it was visible from the road for hours as I rode across the state. Although some of the lava fields are smooth and look like petrified ocean swells, most are broken up into jagged chunks of sharp cinders. Their appearance reminded me of the old cinder running tracks before the days of artificial running surfaces. Many of us old track athletes still carry those cinders embedded in our knees and elbows.

Craters of the Moon National Monument, Arco, Idaho

A seven-mile loop-road through the national monument, with numerous pull-offs, took me by a variety of fascinating points of interest. At the Inferno Cone, I hiked up a six-hundred-foot cinder butte that looked out over ten miles of lava plains. It was a wow-worthy spectacle.

The eruptions are predicted to happen again a thousand years from now. Those interested in seeing the area in its present state ought to make their plans now.

The next day, after visiting the Twin Falls area, I departed Idaho, and crossing into Nevada at Jackpot. It's a tiny town but featured three big casinos. I had gotten in late because of my sightseeing, so I looked for a cheap motel. Alas, all I could find were the big casinos. They looked expensive, but when I inquired I found they were only forty dollars a night. They want to get you there for the gambling, I supposed.

I'd had no communication with Brian since the start of his run but assumed it was because of poor cell coverage. However, riding into Reno, I contacted AT&T and, after a sixty-minute discussion, discovered my cell phone had a problem. I received detailed instructions over the phone on how to disassemble the device, remove its battery, and put it all back together. Hooray, I got bars! It was working! Within the next few minutes, I talked to Michele, to Brian's mother, Sherry Stark, and then to Brian himself.

At Brian's current pace, he said, he would finish in eleven days, not the ten he had initially hoped. Brian was running while we talked and sounded upbeat and optimistic about his completion of the endeavor. We talked about several possible meeting places over the next couple of days.

His mother had received a phone call from Brian the night before and shared his comments in an e-mail she sent to family members.

I was thrilled to get a phone call from Brian last night. He was getting ready to sleep in a real bed. He said he saw his first paved road in six days, yesterday. He told me that his run that day was really challenging. He climbed to 12,000 feet and got into snow. That, however, should be his last mountain, he said.

His run through a slot canyon had been brutal. It had neck-high nettles, thorns, and thick brush. He got cut up pretty badly. The canyon walls were too steep to climb out. After several hours, he was eventually able to climb to the top of the canyon. He said his friend, Hewett, (an EMT) did a good job of cleaning his cuts and scratches, since nettles can leave a nasty sting behind. Brian said at one point he ran alongside some wild mustangs. He also saw a herd of antelope and elk, and in one day crossed twenty-three waist-deep streams. He thinks he has 280 miles remaining and expects to finish on a Monday, rather than Sunday.

He said the support team has been incredible. They drive about three-hundred miles every day to reach places to rendezvous with him. When he arrives at the end of the day, they have his tent set up and freeze dried meals prepared. They drive hundreds of miles to get around mountains and to get to places where they can give him water and sustenance. Last night he was sleeping in a motel bed. Hewett (his EMT friend) was going to sleep on the floor, and the support team was camping at a nearby campground. Brian said the team's vehicles have had four flat tires so far. They had to drive 300 miles to repair the latest flat.

    I rode to Fallon, Nevada, close to where Brian would pass the next day. Time for a motel once again. I can't remember ever being so filthy. After I coated myself with skin moisturizer and sunscreen, my shirts become black around the collar and cuffs. After sleeping on the ground the last few days, I was covered with a film of dust and dirt. My wrinkles were a collection of western soil samples. I don't mean to gross anyone out, but blowing my nose was an archeological dig. I needed a shower, a laundry, and a wash for my bug-enameled motorcycle.

    In Fallon I discovered Brian was also in town, spending the night at an Econo Lodge with his friend Hewett. We met for dinner. As stated in Sherry's email, the challenges had been horrific. Mountain trails that were supposed to exist simply weren't there. Hewett's training as an EMT kept Brian in fair shape, but cuts and scratches on his legs looked like routes of the Paris underground train system. Ted and Trevor are remarkable. While I was visiting with Brian and Hewett, those two were still on the trail, marking it for the next day. Flat tires had become a daily dilemma. Often both vehicles would have flats, and Trevor had to switch wheels around to get one car with four inflated tires, so he could drive 175 miles to fix the flats.

Here's a sixty-four-dollar question for you: what is the C.O.C. & P.P.? Give up? The answer is the Central Overland California and Pikes Peak Express Company. That's the official name of the Pony Express company. Brian is running on much of its route.

In 1860 Russell, Majors and Waddell, an experienced cargo and passenger hauling company, accepted a $600,000 contract from the government to establish a 2,800-mile route overland, guaranteeing that mail could be delivered in twenty-five days. They had sixty-seven days to determine the road, establish 147 stations, hire one hundred riders, no one more than a hundred twenty pounds, for fifty dollars per month, and buy five hundred horses. Each day a rider covered seventy-five to one hundred miles and changed horses eight to ten times. In today's money, it cost eighty-five dollars to send a half-ounce letter by Pony Express. The Pony Express lasted only nineteen months but has fascinated Americans ever since the first riders hit leather in 1860.

Brian would follow the Pony Express route south of Fallon for about twenty-seven miles before crossing Route 95. It would be an ideal opportunity for his fan club to see him in action.

Sherry flew into Reno the night before. Because of her commitments back in Indiana, she could not stay the extra day to see her son's finish at Lake Tahoe. So seeing him crossing Route 95 would be special.

Brian was expected to arrive about noon. I started searching for the crossing spot on the highway about 10:30. I had seen numerous Pony Express trail signs in my travels and expected this spot was also marked. Not so. Those signs are treasured collector's items and frequently disappear. After riding past where the crossing had to be, I turned back and found Ted Oxborrow on his mountain bike, waiting for his son Trevor with the SUV. Trevor arrived within minutes, and the two set up the sun shelter and put out water and nutrients, as they have been doing every ten miles all across Nevada. They are totally professional, organized, and dedicated to supporting Brian. Next to arrive was Hewett in his jeep with the medical supplies.

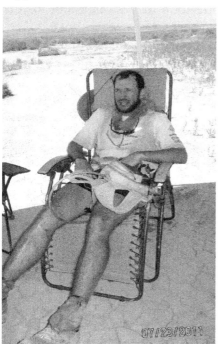

Brian was carrying a walkie-talkie. He checked in, saying he was about thirty minutes away.

But where was Sherry? I couldn't begin to imagine her disappointment if, after flying out here from Indiana and not being able to see him finish, she might now miss seeing him on the run. Reached by cell phone, she said she was on her way and would be there in twenty minutes. It would be close.

Twenty-five miles done, twenty-five more to go this day

Sherry arrived just minutes before Brian came jogging and shouting up the trail, acting like he had been on an errand for a neighbor, rather than running more than twenty-seven miles that morning. Mom gave her youngest a tearful hug. Hewett doused him with cold water, wrapped cold towels around his head, and administered the vitamins. The three of us, Brian, Sherry, and I, sat in camp chairs under the sun awning, admiring Brian's scarred and scabbed legs and hearing his tales of survival. Ye gods! Rattlesnakes, snow passes, aspen groves so thick you had to squeeze between the trunks, bushwhacking, six thousand-foot climbs in four miles, blisters on blisters, and swollen feet. Is he having any fun yet?

After an hour's rest, our runner strapped on his survival belt of water and communication gear and was off chasing the sun in ninety-four-degree heat. Trevor had set out earlier on his mountain bike to set up Brian's next hydration station. Ted took the SUV on a hundred-mile loop to the other side of the mountain, twenty miles from here. Hewett followed Ted. Sherry went north to find accommodations for the night, so she could see Brian again the next day. And I planned to head for a campground at Fort Churchill, another crossing point for the long-distance runner. Before I left, I watched Brian lumbering off up the trail. He says it takes several minutes after a rest to get his body loosened up again. How does he do it — run the equivalent of one marathon, rest briefly, then run another? And do that day after day?

Virginia City, Nevada

Brian is amazing! Instead of wearing down after six or seven days, he seems to get stronger. At the start he was running forty to forty-two miles each day; now he was covering forty-eight to fifty-two miles. The support team members just shook their heads in disbelief. And when he finishes each day, they say, he does a little victory dance with a few enthusiastic war whoops.

What a machine he is!

The afternoon of the next day, Brian was scheduled to do a TV interview on the trail near Reno. To spend the morning until then, I rode up to Virginia City. It's a colorful town in an amusement-park sort of way. Lots of colorful attractions with shops hawking their souvenirs. The real Virginia City of the 1860s became known as the richest place on earth, thanks to the discovery of gold in 1859 in Six Mile Canyon. Almost overnight this town went from a tent city to a city of 30,000 residents, two newspapers, five police precincts, competing fire companies, schools, churches, a Shakespearean theater, opium dens, saloons, and a thriving red-light district. A devastating fire destroyed over two thousand structures in 1875, but the town rebuilt itself in just one year.

I did a quick walk-around of the town, then departed for the media event with the TV crew. Leaving Virginia City, I rode down Six Mile Canyon, an unbelievably steep, twisting canyon road that Brian would be running up later that afternoon. Some sections were so steep, I swear you could stick your hands out and crawl up on all fours.

The television truck was set up in a dusty section of the desert from which Brian had just emerged. He handled himself beautifully in the interview. He mentioned all the causes the run was promoting, to the beaming delight of the Nevada State Parks director, standing off to the side with the American Discovery Trail Coordinator, who organized Brian's run. He then credited his individualism and courageous independence to his parents, Jim and Sherry Stark, who left him behind at a rest stop when he was seven years old and didn't realize he was missing until they were three states away. The groan from his mom was just barely audible in the interview broadcast.

"My parents drove off and left me in New Jersey when I was little."

# # #

Chapter Thirty-Nine

The Finish

I checked into the Cal-Neva Hotel at Lake Tahoe to await Brian's finish. His run today was only thirty miles, and Brian called to say he would arrive at the finish line about 1:30. I spent a leisurely morning and changed hotel rooms to one with two beds, so Brian could stay the night before flying back to Tucson the next day. Later I took a short ride into California and took a few pictures of beautiful Lake Tahoe before returning to the hotel to await Brian's arrival.

As Brian ran down the last hill, two high school cross-country runners approached him and said, "Someone said there is a guy out here running across Nevada. Is that you? Can we run along with you?" They dropped off after two miles, but not before Ted got some good video of the three of them for the documentary he is filming. When Brian was four miles out, he could look out over Lake Tahoe and spot the Cal-Neva Hotel. He sprinted the last four miles running a sub-eight-minute pace.

The finish, Nevada/California state line

Brian crossed the Nevada/California State line in eleven days, seven hours, and twenty-eight minutes from the time he started, becoming the first person to run the American Discovery Trail in its entirety across Nevada. He crossed fifteen mountain ranges, for a total climb of 34,000 feet. Mount Everest is only 29,000 feet, so how's that for a comparison?

What an extraordinary accomplishment. The gauntlet was now down for anyone wishing to beat Brian's record. And that was the point of Ted Oxborrow's invitation, to promote adventure tours in Nevada. Whether running, biking, or hiking, come to Nevada and challenge the beautiful mountain topography. Videos, press releases, and magazine articles were planned to tell of Brian's run.

That night Brian and I spent a delightful evening, starting at the hotel's dinner buffet. For an interesting sight, you ought to see someone who has been burning an extra five thousand calories a day load his plate at an all-you-can-eat buffet table. Not once but three times. When I thought he must be ready to explode, Brian pushed back from the table and announced, "All right, time for dessert!" Amazing.

We finished the evening with a visit to the casino. I never had much interest in games of chance, so know very little about the games. Brian, however, was surprisingly good at blackjack, and won nearly one hundred dollars.

The following morning, Brian, after a visit to the breakfast buffet, caught the hotel's shuttle to the airport for his flight home to Tucson. I loaded up, turned my motorcycle 180 degrees, and headed east on Highway 80, back across Nevada.

I spent the entire next day riding across Nevada. It wasn't until 6:00 p.m. that I finally reached its eastern border. What a wide state! Can you believe that, Brian? Oh yeah, you can, sorry.

When I departed Nevada and entered Utah, I rode along the Bonneville Salt Flats. The explorer Joe Walker, mapping this area in 1833, named the flats after his boss, Captain Ben Bonneville. The flats originated from the Great Salt Lake, which flooded over a third of Utah 32,000 years ago. When the sea receded, it left a crust of potash and halite (table salt). The crust at the center of the flats is five feet thick, but at the edges may be only an inch or two.

Bonneville Sail Flats, Utah, 2011

As I rode along the highway, I could see the flats came right to the edge of the road and extended for miles, looking like a vast mall parking lot. The thought occurred to me to cut over onto the surface and see how close I could come to the 150 mph limit on my Goldwing's speedometer. Car tracks leading from a highway to the flats indicated I was not the first to have that idea. However, I also noticed that there were several tracks that had broken through the thin crust at the edge and must have resulted in a towing charge, pulling the cars out of the brine back onto the highway.

I slugged it out the rest of the day, enjoying the Utah sunshine, with its ninety-degree temperatures. Once I got to Salt Lake City, I turned southeast toward Colorado with plans to follow a route my oldest son, Eric, had recommended. He attended a music conference in Aspen and had several suggestions of sights I would enjoy.

When I approached Provo, I gained elevation and had to stop to put on my motorcycle jacket as the temperatures dropped into the sixties. It was another wonderful day, but for the first time, I was feeling extreme fatigue. It wasn't drowsiness, but the feeling of exhaustion.

My friend Dave Baer, a fellow former Navy pilot, said, "Stark, you should've been a Marine." Well, I was beginning to feel like one. I slept in the dirt in many of my campgrounds, and I've never been as grungy as I am after riding through these desert climates. It's often two days between showers.

I found a pull-off rest area. Although there was an outhouse and picnic tables, the high sun allowed little shade. Pulling a garbage bag out of my pack, I spread it under a low bush and lay down in the sand and sacked out for half an hour. I awoke with bugs in my ears and brambles in my butt, but refreshed and ready to ride. Semper Fi!

I crossed the Colorado line and stopped at Grand Junction. I found a motel with laundry facilities (maybe the last I would need), internet, and in-room conveniences. Ah, the good life!

Independence Pass. Colorado

What a spectacular day the next turned out to be. Right from the get-go as I left Grand Junction, I rode through canyons that in the early morning light, were illuminated like Thomas Kinkade paintings. I stopped at least three times, trying to catch their brilliant images with my camera. Eric suggested I turn off at Glenwood Springs and take Route 82 to Aspen. What a scenic, twisting road that was! At Aspen I got a map at the visitor center and walked the downtown shopping area with my eyes peeled for celebrities. All the designer shops were there: Louis Vuitton, Prada, Gucci, and many names with which I'm not familiar. Real estate offices had pictures in the windows displaying their bargain specials, but $9,500,000 for a home built of logs seemed a bit much. I was wearing my last clean shirt, just in case I bumped into Nicole Kidman. I had a couple of shoppers snap my picture just to prove I was doing Aspen.

A visitor center guide directed me to ride Independence Pass. Holy Toledo, what an experience that was! The pass winds along the mountainside to an elevation of 12,093 feet. Often there are no guard rails, and the road becomes so narrow that oncoming cars must wait at a wide place in the road to get by. I decided this would be a good place to make a video recording. The problem was, some of the maneuvering required a ten mph or slower speed, and I couldn't get my cruise control to kick in below thirty-five mph. So, with one hand on the throttle and the other holding the camera, I filmed several minutes of that unbelievable ride. Two times during the climb I suddenly had to drop the camera, secured by a strap around my neck, and grab for the handlebars. Looking at the video, you see some blurred images and my shirt at two points in the film.

At the summit, which was also the Continental Divide at 12,000 feet, I was well into the snowfields. I dismounted for several pictures. I'd covered only thirty miles in the previous hour, but those were once-in-a-lifetime views. What a remarkable state Colorado is! On the downward plunge I was able to sweep and roll like a torpedo bomber on a run at the Battle of Midway. I was leaning into my turns, first one way, then another, with a grin on my face that stretched from ear to ear. It was fun to consider that just thirteen years earlier, Brian's five-thousand-mile run across America also brought him to this very summit at Independence Pass. We are both weaving our own fabric of adventure, with occasional overlapping threads.

Hours later, after passing through Denver and approaching Strasburg, Colorado, I looked for a motel. Highway signs said accommodations were available, but it took a search to find the inn, an older home with six rooms to rent. The door was locked, but a note invited me to ring a bell.

"Do you have internet?" I asked.

"No, but we're thinking about it."

"How about ice?"

"No, we're out, but I'll go downtown and buy you some."

"Sounds good. Sign me in."

I sat in the inn's backyard, fixed a drink, used my camp stove to prepare a freeze-dried dinner, and worked on my journal. What could be better than that?

I got up early, with plans to load my motorcycle for an early departure, then grab a quick breakfast at the café, the only restaurant on the main drag. Loading the motorcycle was its usual arduous affair, carrying bags, cooler, map cases, and electronics, and getting all into their designated places. The front door of the inn was locked, and although I was given a key besides my room key, repeated trips were complicated by the extra step of unlocking the front door each time. Finally I was loaded and made my final inspection of the room to make sure I had forgotten nothing.

The opportunity to overlook something has been great on this trip. Whenever I'm spread out over a campsite or a motel room, I have a lot of stuff, some of which would be a sad loss if overlooked: cell phone, still camera, video camera, charging cords for all the above, my glasses, computer, flash drives, camp stove, cook kit ... well, you get the idea. So far, my only problem had been sunglasses; I was on my fourth pair. I misplaced my regular glasses once at the Fort Churchill State Park but then found them in the dirt in the front of my tent. That was close. On previous trips I'd always brought a backup pair, but this time I'd forgotten.

So I checked the room twice, which wasn't easy, because it had a lot of frilly surface coverings like doilies and flowered bedspreads, which camouflaged the surfaces. Completing my inspection, I left my keys on the front lobby desk and got ready to ride. And soon as I put on my helmet, it hit me. Where were my hearing aids? I hadn't put them in my ears, and they weren't in the case I carry for that purpose. They must still be on the table by the bed, but now I'm locked out. The inn had a doorbell, but after ten minutes of ringing, I surmised the owner did not live on the premises, and with its being 6:15 a.m., who knew when he would arrive? Expletive! Double expletive! There went my early departure.

I had my cell phone, but nowhere on the building or on my receipt was a listed telephone number. I started poking around the property. Behind the inn was a dilapidated trailer, but it looked lived in. I knocked several times on the door, holding my breath until I was finally answered by someone obviously aroused from sleep, and yes, it was the owner. I explained my dilemma, and he led me into the back entrance of the inn, through a door I hadn't noticed that led to my room's bathroom. There they were, my hearing aids, on the bedside table. I thanked him and exited by the front door. I don't think the man said one word during the entire episode.

Mark Twain boyhood home, Hannibal, Missouri

Whenever someone expresses the cliché "It's a small world," you can rest assured they never crossed Kansas on a motorcycle. Man, that's a big state! I spent the day in transit and still didn't get it done. I spent the night in Marysville, Kansas.

After I arrived the following day in Hannibal, Missouri, I decided I needed a break and visited Mark Twain's boyhood home. Samuel Clemens, alias Mark Twain, was actually born in Florida, Missouri, in 1835 and moved to Hannibal with his family as a four-year-old. Sam's father also brought six slaves with him, whom he had inherited from a deceased family member. His father was elected justice of the peace in Hannibal but, its being such a low-paying office, tried other trades with little success. The slaves were sold one by one to finance living expenses. Samuel grew up in Hannibal and used his boyhood experiences as background for *Adventures of Huckleberry Finn*, years later, in 1876. The family's last slave, Uncle Daniel, became Tom Sawyer's fictional pal Jim. Tom Sawyer was based on Sam's scallywag friend Tom Blankenship.

Clemens made a return visit to his boyhood home in 1905, just five years before his death. The museum has mannequins of Sam and comments he made during his visit to the old homestead: "Reminds me of Mom and Dad" and "I recall my old friends."

Samuel Clemens is remembered for his many quotations, perhaps the most memorable being "It is better to keep your mouth shut and appear stupid than to open it and remove all doubt." My personal favorite is "I don't know anything that mars good literature so completely as too much truth." If you substitute "a good story" for "literature," you have a motto my mother used to live by.

War wounds, Springfield, Illinois

I spent my last night on the road in a campground in Springfield, Illinois. I set up my tent in a shady grove after stripping down to shorts and a T-shirt. As I was updating my journal, two young campers, Tray and Kaiser, came calling and wanted to know if I would like s'mores. Oh yeah. Marshmallow, chocolate, and Graham crackers are an important part of my diet. Bring it on, kiddos. Mmmm, good, and of course I had to tell them all about Brian.

Then Theriol came over, a young black girl of maybe five, who wanted a push on the merry-go-round.

"Theriol, that's such a pretty name. How do you spell it?"

"I don't know."

I got her spinning so fast that it looked like fun. When I tried to hop on, a hand-bar hit me in the lower leg and opened a gash. Not enough for stitches, but bloodletting nevertheless. What's a good road trip without a few war wounds!

So, what is it about this motorcycling thing that I find so compelling? Riding a motorcycle is like pointing your front wheel into the screen of an IMAX, 3-D theater, with surround sound, surround smell, and surround wind. You can't stop swiveling your head to see it all, 180 degrees all around you. All of this with your feet a mere five inches above the concrete highway. You marvel, you gasp, you catch your breath, and you find yourself grinning like a fool. You can witness the same scenes from a car, but there you are enclosed, separated, and not part of it. I've never been drowsy on a motorcycle. I've been tired for sure — weary, bone-aching, neck-throbbing exhausted - but I've never been sleepy. My senses are constantly stimulated. Even the flat-out nothing of Kansas amazed me with its vastness, and I couldn't stop marveling at it.

Every day of this trip was exhilarating. Every day there were extraordinary sights to see that just made me want to shout out in exuberance. What excitement, what thrills - and thank you, Lord, I'm finishing this trip all in one piece.

> *"Life should not be a journey to the grave with the intention of arriving safely in a pretty and well preserved body, but rather to skid in broadside in a cloud of smoke, thoroughly used up, totally worn out, and loudly proclaiming, 'Wow! What a Ride!'"*
>
> Hunter S. Thompson

\# \# \#

Made in the USA
Monee, IL
18 June 2020